COGNITIVE NEUROSCIENCE, 2010, 1 (3), 153–154

Introduction

The cognitive neuroscience of consciousness

Geraint Rees[1] and Anil K. Seth[2]

[1]University College London, London, UK
[2]University of Sussex, Brighton, UK

The past decade has seen an explosion of interest in the neural mechanisms underlying consciousness. Experimental approaches from cognitive neuroscience that emphasize converging evidence from multiple methodologies have changed our understanding of how conscious mental states are associated with patterns of brain activity. In this special issue of *Cognitive Neuroscience*, we bring together five new empirical contributions to this literature plus a new theoretical discussion paper and associated peer commentaries.

To understand the relationship between conscious mental states and brain activity, we must make progress in three distinct areas. To be conscious is to be awake (rather than in a dreamless sleep, or unconscious) and so we must understand the neural mechanisms associated with changes in level of consciousness. But when we are awake, our conscious states are individuated by their content, which has a particular subjective feel. We therefore need to understand the neural underpinnings of conscious content, and how such representations in the brain are distinguished from merely unconscious processing. And finally humans (and possibly some other animals) are self-aware and able to make introspective judgments about their perception and action. Understanding the neural correlates of such metacognitive ability is also required.

In this special issue, exciting new empirical data are presented in each of these areas. Bruno et al. (this issue) focus on the challenging problem of assessing patients in a minimally conscious or persistent vegetative state. These disorders are difficult to characterize clinically and there has been much recent interest in using functional neuroimaging to investigate possible residual cognition and consciousness. Bruno and colleagues review the relevant work, and present new data collected in the resting state that may prove helpful in diagnostic and prognostic classification.

The level of consciousness does not only change following brain injury, but follows a natural pattern in healthy animals of sleep/wake cycles. Massimini and colleagues (this issue) focus on the consequences of such cycles for cortical processing, using transcranial magnetic stimulation in combination with high-density EEG to show striking changes in cortical reactivity and effective connectivity during rapid eye movement (REM) sleep compared to non-REM sleep; notably, the changes observed in REM sleep are similar to those observed in wakefulness. These findings suggest that measuring cortical reactivity and effective connectivity may help evaluate whether an individual is capable of having conscious experiences, and so have wider application in brain injured patients.

Brain injury can also cause selective impairment to the contents of consciousness. Spatial neglect commonly follows right parietal injury and results in difficulty noticing or responding to objects in the contralateral visual field. In this issue, Eramudugolla,

Correspondence should be addressed to: Geraint Rees, UCL Institute of Cognitive Neuroscience, University College London, 17 Queen Square, London WC1N 3AR, UK. E-mail: g.rees@ucl.ac.uk

www.psypress.com/cognitiveneuroscience
DOI: 10.1080/17588928.2010.503602

Mattingley, and Driver report a single case where awareness of contralateral stimuli critically depends on biases in figure–ground segregation induced by changing the relative salience of figure and ground.

Two studies explicitly examine self-awareness, but from different perspectives. Menzer and colleagues (this issue) focus on the sense of agency that we experience in linking the sound of our footsteps to our own actions. By manipulating the delay between walking and hearing the sound of footsteps, they examine the temporal dependency of this sense of self-agency. In contrast, Rounis, Maniscalco, Rothwell, Passingham, and Lau (this issue) examine how humans make introspective (metacognitive) judgments about their accuracy in a simple visual discrimination task. Transcranial magnetic stimulation over dorsolateral prefrontal cortex impairs such metacognition while leaving actual discrimination unaffected, indicating a role for this structure in making metacognitive judgements.

All empirical work ultimately requires a theoretical framework, and in this special issue Victor Lamme provides a thoughtful and provocative discussion paper attracting vigorous peer commentary. Lamme presents an extended argument that the phenomenality of conscious experience is explained by recurrent neural activity, contrasting this position with the notion that consciousness is associated with activity in an extended fronto-parietal network. Critically, Lamme argues that consideration of recurrent processing as a fundamental explanation for phenomenality allows us to take seriously the notion advanced by philosophers such as Ned Block (e.g., Block, 2005) that we can have phenomenal states that are unreportable. On Lamme's account, phenomenality arises through recurrent processing, while it is reportability that requires involvement of a frontoparietal network.

Taken together, the work reported in this themed issue of *Cognitive Neuroscience* demonstrates the empirical vigour and ingenuity that characterize this field, together with the increasing theoretical sophistication and debate surrounding elaborated theories of visual awareness. Such interaction between theoretical debate and empirical work is testimony to the rapidly growing maturity of this work on the cognitive neuroscience of consciousness.

REFERENCE

Block, N. (2005). Two neural correlates of consciousness. *Trends in Cognitive Sciences, 9*, 46–52.

COGNITIVE NEUROSCIENCE, 2010, 1 (3), 155–164

Biased figure–ground assignment affects conscious object recognition in spatial neglect

Ranmalee Eramudugolla[1], Jon Driver[2], and Jason B. Mattingley[1]

[1]University of Queensland, St Lucia, Australia
[2]University College London, London, UK

Unilateral spatial neglect is a disorder of attention and spatial representation, in which early visual processes such as figure–ground segmentation have been assumed to be largely intact. There is evidence, however, that the spatial attention bias underlying neglect can bias the segmentation of a figural region from its background. Relatively few studies have explicitly examined the effect of spatial neglect on processing the figures that result from such scene segmentation. Here, we show that a neglect patient's bias in figure–ground segmentation directly influences his conscious recognition of these figures. By varying the relative salience of figural and background regions in static, two-dimensional displays, we show that competition between elements in such displays can modulate a neglect patient's ability to recognise parsed figures in a scene. The findings provide insight into the interaction between scene segmentation, explicit object recognition, and attention.

Keywords: Attention; Figure-ground segmentation; Object recognition; Parietal lobe; Spatial neglect.

INTRODUCTION

Spatial neglect is a complex disorder of attention and spatial behaviour that follows unilateral brain damage, predominantly of the right hemisphere, manifesting as reduced exploration and conscious awareness of stimuli presented to the side of space opposite (or contralesional) to the damaged hemisphere (Driver & Mattingley, 1998; Karnath, Niemeier, & Dichgans, 1998; Mesulam, 1999; Stone, Halligan, & Greenwood, 1993). These symptoms occur despite intact primary sensory and motor functions, and instead reflect a range of "higher-level" deficits in attentional orienting (Corbetta and Shulman, 2002; Driver and Vuilleumier, 2001), representation of contralesional space (Bisiach and Luzzati, 1978; Halligan, Fink, Marshall, & Vallar, 2003) and biased coordinate frames for exploratory behaviour (Karnath et al., 1998). The neglect syndrome has provided important insights into the relationship between different levels of stimulus processing in the brain and the conscious perception of these stimuli (Driver and Vuilleumier, 2001). In the present study we examined object recognition processes in neglect, focusing specifically on how an underlying bias in figure–ground segmentation can lead to failures in conscious recognition of otherwise familiar shapes.

Correspondence should be addressed to: Jason B. Mattingley, Queensland Brain Institute, The University of Queensland, Australia, 4072. E-mail: j.mattingley@uq.edu.au

We thank RH for his participation, and Tom Manly for his assistance with testing.

RE was supported by an NH&MRC Project Grant awarded to JBM and RE. JD was supported by the Wellcome Trust, and by a Royal Society Anniversary Research Professorship.

www.psypress.com/cognitiveneuroscience   DOI: 10.1080/17588921003605376

Many previous studies of the neural correlates of conscious perception in healthy individuals have had participants view rivalrous, bistable, or otherwise ambiguous images, such as those exemplified by Rubin's famous "faces–vase" display (for a review, see Kim & Blake, 2005). In such situations, the display remains unchanged while the viewer's conscious experience switches between two possible visual interpretations. In the current study, we took an analogous approach. We constructed ambiguous visual displays from familiar objects (such as the table lamp in Figure 1A). For each object we created two otherwise identical, half-object images in which a central contour defined a familiar shape on one side, and a meaningless background region on the other (Figures 1B and 1C). Over a series of experiments, we compared conscious recognition of the familiar shape by varying the orientation of the central contour, while holding all other aspects of the display constant.

Studies of figure–ground segmentation in neglect patients suggest that neglect arises at a stage after the perceptual parsing of visual input (Driver et al., 1992; Marshall & Halligan, 1994; Peterson, Gerhardstein, Mennemeier, & Rapcsak, 1998). There is also evidence that the identity of neglected stimuli can have implicit effects on neglect patients' performance (see Driver & Vuilleumier, 2001). These intact perceptual and object identification processes make neglect a useful context in which to examine the relationship between attention, scene segmentation, and object recognition.

Within a visual scene, regions that are smaller in size, more convex, brighter, or symmetrical tend to be perceived as figures rather than background. Driver et al. (1992) presented a neglect patient with figure–ground displays in which a single vertical contour divided a region of space into a small bright area on one side and a dimmer, larger region on the other. In this manner, low-level cues were used to bias figural assignment to the smaller, brighter region. When asked to attend only to the contour and judge whether it matched a probe contour, the patient showed a striking bias: He responded accurately when the figure was on the left of the display (with the contour defining the right edge of the figural region) and poorly when the figure was on the right of the display (with the contour defining the left edge of the figural region). Although superficially counterintuitive, this result indicates that the spatial bias in neglect operates on objects *following* their segmentation from a scene, rather than on unstructured regions of space *per se*.

Object-centered manifestations of neglect are well known in patients' drawing and copying, where patients depict the right edge of an object but typically fail to draw its left edge (e.g., Driver and Halligan, 1991; Behrmann & Moscovich, 1994; Gainotti and Tiacci, 1970). This impairment in consciously representing the contralesional edge of objects also influences the way patients assign edges to shapes and thereby segment regions of space in a visual scene. Specifically, patients demonstrate a bias toward assigning an edge or contour to the region of space on the left of the contour, making that contour the right edge of the shape (the configuration depicted in Figure 1C). For example, Marshall and Halligan (1994) found that a neglect patient could accurately depict the central contour of an ambiguous figure–ground display if asked to copy it as belonging to the edge of the left-sided figure, but not if asked to copy the same contour as the edge of the right-sided figure. Thus, although figure–ground segmentation may occur prior to the allocation of focused attention (Kimchi & Peterson, 2008), and in an obligatory fashion (Baylis & Driver, 1996), biased spatial attention can nevertheless act in a top-down manner to influence figural assignment. This is further supported by data from neurologically normal individuals showing that attentional cueing can bias their figure–ground segmentation (Baylis & Driver, 1996; Vecera, Flevaris, & Filapek, 2004).

In the patient studies cited above, the intended figural region was disambiguated from the background, either explicitly by the examiner (Marshall & Halligan, 1994) or by using low-level cues such as size and brightness (Driver, Baylis, & Rafal, 1992). Thus the direction of spontaneous edge assignment was either pre-empted or heavily biased toward the intended figure. Moreover, these previous studies required patients to focus their attention on the local features of edges themselves, rather than on the figural and background regions thus defined.

What are the implications of biased figure–ground segmentation for the perception of meaningful figures? Most previous patient studies have employed only ran-

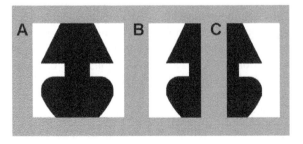

Figure 1. Example stimuli used to test figure–ground assignment in neglect. (A) Whole-object display of a table lamp. (B) Half-object display of a table lamp, in which the central contour divides a meaningful shape on the right side from a meaningless background region on the left. (C) Same half-object display as in (B), but with the meaningful shape on the left of the central contour.

dom contours and nonsense shapes, rather than edges and shapes depicting real (i.e., meaningful) objects. There is evidence from normal observers, however, that stored shape representations can exert a top-down effect on figure–ground segmentation (Peterson & Gibson, 1994). A similar finding was made for patients with unilateral left- or right-hemisphere lesions (Peterson et al., 1998). In this study, Peterson et al. (1998) used ambiguous figure–ground displays with a central contour in which low-level cues to figural assignment were minimized. Half the displays included a shape that depicted a meaningful, familiar object such as the table lamp shown in Figures 1B and 1C. Patients were more likely to perceive the ipsilesional side of the displays as the figure when these regions represented meaningful objects than when both sides of the contour formed a meaningless shaped region. Peterson et al. (1998) suggested that even though attention is reduced for contralesional edges, object recognition processes can help segment these regions prior to the allocation of focused attention. Interestingly, their patients were also somewhat poorer at *identifying* meaningful objects on the ipsilesional than on the contralesional side of the central contour, though this trend was not statistically significant.

In the present paper, we demonstrate that conscious object recognition is reliably impaired for regions on the ipsilesional (right) side of a central contour in a right hemisphere patient with left-sided neglect, despite the patient's preserved ability to identify the same objects when they are shown on the left (contralesional) side. We examined whether this impairment in identifying objects to the right of the central contour is apparent even when the patient is provided with the correct figural assignment. The findings indicate that a bias in object recognition is evident both when the patient spontaneously chooses the right side of the display as the figure, and when he is explicitly cued to attend to the right in order to identify a familiar object on that side. Finally, we hypothesized that this bias in conscious object identification in neglect may be influenced by the degree of competition between figural and background regions on the two sides of the central contour (Peterson & Salvagio, 2008; Peterson & Skow, 2008). In support of this, we found that explicit identification of an ipsilesional shape is significantly improved when the salience of the shaped region on the contralesional side of the central contour is reduced.

EXPERIMENT 1

In Experiment 1, a patient with a right hemisphere lesion and left spatial neglect was presented with ambiguous figure–ground displays containing a central contour. In each display, one of the two regions separated by this critical contour, or edge, represented a real object. We investigated whether there was any lateralized bias in this patient's spontaneous assignment of figural status to one of the two regions of the display, and in his ability to identify objects that were correctly segmented.

Method

Case report

At the time of testing, patient RH was a 66-year-old man who had suffered a right temporoparietal stroke one year previously (see Figure 2). At the time of testing, RH demonstrated left neglect and left-sided extinction on several standard clinical measures (Table 1). In scene copying he omitted items toward the left of the scene, but correctly completed items toward the right. When copying a clock he omitted digits on the left of the clockface, and in drawing a person from memory he omitted the left arm. There was no evidence of visual object agnosia or other disturbances of visual perception on informal testing.

Materials

The stimuli were black and white, two-tone images printed onto sheets of white A4 paper, trimmed to the sizes specified below, and then laminated.

Practice stimuli. The patient was initially familiarized with the task of choosing a figure from a background by presenting him with a set of stimuli comprising small geometric (*n* = 4) and nonsense (*n* = 4) shapes located in the center of a larger rectangular background. Each of the eight shapes was presented twice, once in black against a white background and once in white against a black background. The task was to report the colour of the object (black or white), thus introducing the patient to this way of communicating figure–ground distinctions, for displays that had unambiguous figure–ground relations, before proceeding to the experimental figure–ground displays.

Whole-object stimuli. A set of stimuli was created in which a meaningful shape was presented in full and in isolation with an abutting background region within a rectangular window (Figure 3A). There were 24 different shapes, which were adapted from Gibson and Peterson (1994) and from the Snodgrass and Vanderwart (1980) set. Each such shape appeared in

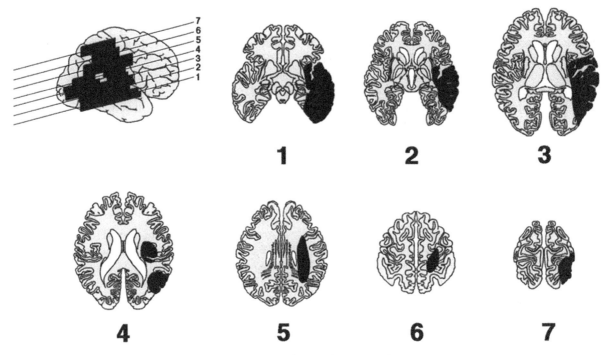

Figure 2.   Lesion reconstruction based on brain CT scan for RH taken two days post-stroke.

TABLE 1
Patient RH's performance on clinical tests of spatial neglect and extinction

	Left	Right	
Albert's lines	0/18	13/18	
Star cancellation	0/27	13/27	
Letter cancellation	0/20	10/20	

	Percentage bias +32.1% (rightward)		
Line bisection			

	Left unilateral	Right unilateral	Bilateral
Auditory extinction	10/10	10/10	3/10
Tactile extinction	10/10	10/10	3/10

	Left	Right	
Fluff test	3/12	11/12	
Wundt-Jastrow illusion	4/20	18/20	

black against a white surround, and vice versa, making a total of 48 different whole-object stimuli.

Experimental stimuli. The experimental stimuli each comprised outlines of the left or right half of a meaningful object (created from the whole-object stimuli; see above) (see Figure 3B). The object appeared either to the left or right of a central contour within a rectangular display. Note that when the object appeared to the left of the central contour, the contour represented the object's right edge, whereas when the object was located to the right of the central contour, the contour marked the left edge of the object. Thus the side of the object (in object-centered coordinates) that was depicted in the display was always opposite to its location relative to the critical dividing edge within the rectangular display. In half the stimuli, the object was black against an abutting white background, and in the other half white against an abutting black background. Thus the 24 different objects yielded 96 figure–ground displays, such that the set of stimuli was internally counterbalanced according to side (left or right) and shade (black or white) of the object from which it originated.

Procedure

All stimulus items were presented in free vision against a uniform gray background. Unless otherwise stated, there was no limit in the viewing duration. Initially, RH was shown the practice stimuli and asked to indicate whether the shapes depicted were black or white. He was then shown, in random order, the whole-object displays. The aim was to prime the patient on the identity of the meaningful objects depicted. During the presentation of whole-object stimuli, RH was asked to indicate first the shade and then the identity of the figure in each stimulus, and feedback was provided for any shapes not correctly

A

B

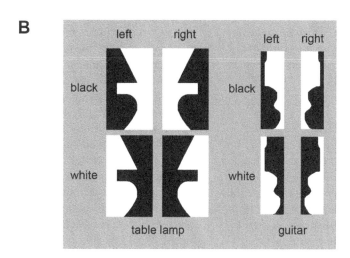

Figure 3.   Example stimuli used in Experiments 1 and 2. (A) Whole-object displays from which the figure–ground stimuli were generated. (B) Half-object displays used as ambiguous figure–ground stimuli. Due to the constraint of matching the areas of adjacent black and white regions for very different shapes, the resulting figure–ground stimuli varied in their vertical (112–170 mm) and horizontal (48–109 mm) extent. Stimuli were presented individually on a uniform gray surface (30 × 40 cm) that was selected to be intermediate in contrast between the black and the white in rectangular figure–ground displays. This gray viewing surface was inclined toward the patient at an angle of approximately 45°. Each figure–ground stimulus was presented in its canonical (upright) orientation, and aligned with the patient's midsagittal axis at a viewing distance of approximately 60 cm.

identified. Following this, RH was presented with the experimental stimuli in random order. For each stimulus, he was asked to indicate the shade of the figure, and then its identity. For the experimental trials no feedback was given on accuracy and RH was encouraged to guess if he was unsure of the shade or identity of a figure.

Results and discussion

During the priming session with whole-object stimuli, RH was at ceiling in choosing the central region as the figure, as would be expected for these displays in which relative size and surroundedness provided salient disambiguating cues to the figure–ground relations. On initial presentation RH identified all but 6 of the objects correctly, and after corrective feedback he was able to identify all 48 objects readily.

For the experimental stimuli, RH showed a striking asymmetry in his figural assignments. Responses were analyzed using two-way chi-squared tests, with Yates' correction where appropriate. When the object appeared on the *left* of the display (i.e., so that the central edge defined the shape's *right* side), RH chose it as the figure on every trial. In contrast, when the same object appeared on the *right* of the display (i.e., so that the critical edge defined the shape's *left* side), RH chose it as the figure on only 27.1% of occasions. Thus he chose the object as the figure more often when it appeared on the left of the display than when it appeared on the right: $\chi^2(1) = 55.1$, $p < .001$. In addition to RH's tendency to choose the left side of the displays as the figure even when the object actually appeared on the right, he also had trouble identifying objects on the right even when he did see them as the figure. Considering only those trials for which a correct figural assignment was made, RH identified 81.3% of objects when they appeared to the left of the critical dividing edge, but only 46.2% of objects when they appeared to the right of the critical edge, $\chi^2(1) = 4.8$, $p < .05$.

These results are consistent with previous reports that neglect patients tend to assign the critical edge of figure–ground stimuli to the region on the contralesional side, so that it defines the right edge of the shaped region on that side (Driver et al., 1992; Peterson et al., 1998). The novel finding of Experiment 1 is that even when the object on the ipsilesional side of the display was spontaneously perceived as the figure, the patient was still significantly poorer at identifying the object than when it appeared on the contralesional side. Thus RH's bias appeared to affect not only figure–ground segmentation, but also the recognition of parsed objects, and this was examined further in Experiment 2.

EXPERIMENT 2

In Experiment 1 we asked RH to choose which side of the experimental stimuli contained a meaningful object, and then to try to identify the shape he selected. Since RH rarely saw the region to the right side of the critical edge in our displays as figural, our measure of his impairment in consciously identifying figures on that side was necessarily based on only a small number of stimulus displays. To circumvent this problem, in Experiment 2 we explicitly directed RH to attend to the side of the display that contained the meaningful shape by cueing him with the shade of the object (black or white) that he had to identify.

Method

The stimuli and procedure for presentation were the same as those described in Experiment 1. Prior to the main experimental task, RH was again asked to identify the set of 48 whole-object displays, which he was able to do without error. RH was then shown each of the experimental stimuli. In an initial *unspeeded* condition, the examiner indicated the correct figural assignment (i.e., which of the two regions within the rectangular display was the shape) by telling RH the shade of the shape he was to identify (either black or white) immediately before the stimulus was displayed. RH was permitted to respond at his leisure, without any explicit time pressure. Unlike Experiment 1, where RH chose the figure himself, we could now score identification performance for the meaningful object on every trial. RH was encouraged to guess if he was unsure of the identity of the object, and no feedback on accuracy of responses was given. In a subsequent

speeded condition, which was conducted a week later, RH was required to identify the cued shape as quickly as possible, and his vocal responses were timed using a stopwatch.

Results and discussion

In the *unspeeded* session RH correctly identified 92% of left-sided objects but only 63% of right-sided objects, $\chi^2(1) = 11.56$, $p < .001$. This result replicates the asymmetry in identification accuracy that was observed in Experiment 1. The same pattern of performance was observed again a week later, when we timed RH's identification responses in the *speeded* condition. RH correctly identified 94% of objects on the left and only 58% of objects on the right, a difference that was again statistically reliable, $\chi^2(1) = 16.53$, $p < .001$. For those stimuli in which the object was correctly identified, the naming latencies also reflected a distinct asymmetry. Figure 4 shows a scatterplot with each point representing the naming latency for those objects that RH identified correctly regardless of the side on which they appeared. The figure clearly shows that the majority of points lie to the right of the positive diagonal, indicating that RH was consistently slower to correctly identify objects on the right side of the display than on the left. This asymmetry was confirmed by comparing RH's mean identification latency for left objects (mean = 2.16 s; $SD = 0.63$ s) vs. right objects (mean = 3.70 s; $SD = 1.73$ s), $t(26) = 4.58$, $p < .001$). Thus, even when RH succeeded in identifying an object on the right side (with the central edge on the left), he was still significantly slower to do so than for the same object on the left (with the central edge on the right).

EXPERIMENT 3

In Experiments 1 and 2, the central dividing contour could always be assigned in two possible directions (left or right), with profound consequences for conscious identification of the shapes thus defined. Models of figure–ground perception suggest that segmentation results from inhibitory competition between regions on either side of a dividing edge (Peterson & Salvagio, 2008; Peterson & Skow, 2008; Sejnowski and Hinton, 1987; Zipser, Lamme, & Schiller, 1996). Data from normal participants indicate that conscious perception of the resulting figural region is enhanced while that of the ground region is suppressed (Peterson & Skow, 2008).We reasoned that in neglect, the presence of a salient background

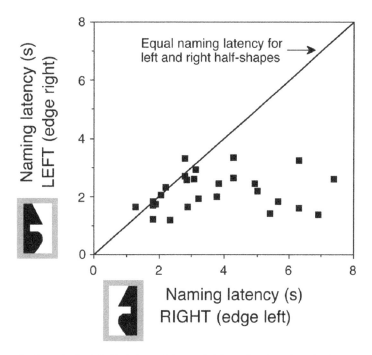

Figure 4. Scatterplot of naming latencies for left-sided shapes (edge on right) as a function of naming latencies for right-sided shapes (edge on left). Each point represents a specific object shape (e.g., table lamp). Note that naming latencies for an object presented as a figure on the right side of the display (edge on the left) were often considerably slower than the naming latency for the same object presented as a figure to the left of the display (edge on the right).

region on the contralesional side of the central contour to the meaningful object might act as a potent competitor during figure–ground segmentation. In a final experiment, we tested the hypothesis that RH's difficulty in identifying objects on the right of the central edge might be due to biased edge assignment in the context of perceptual competition between abutting figural and background regions of the display. If this hypothesis is correct, then reducing the salience of the background on the left side of the central contour should improve identification of objects on the right. To achieve this, we simply removed the competing background region for half of the displays, so that the figural region always appeared as an entirely enclosed shape surrounded by a uniform gray background (see Figure 5A).

Method

Stimuli

We used the same stimuli as in the previous experiments, except that they were now presented on the integrated LCD monitor of an Apple Macintosh laptop computer. Each experimental stimulus appeared in two different configurations. In the "competing" configuration, the stimuli were identi-

cal to the black-and-white rectangular displays used in Experiments 1 and 2. In the "noncompeting" configuration, the same objects were presented without the abutting high-contrast region that completed the rectangle as background, so that the figure was now entirely surrounded by a uniform gray field that was intermediate in contrast between black and white (see Figure 5A). There were 24 different objects, each presented on the left or right side in either black or white. The stimuli were displayed in both the competing and noncompeting configurations, yielding a total of 192 different stimuli, which were presented randomly intermingled.

Procedure

RH viewed the screen from a distance of approximately 60 cm, so that the displays subtended the same visual angles as those used in Experiments 1 and 2. At the commencement of each trial RH fixated a cross that appeared in the center of a uniform gray field. After a delay, the cross disappeared and was replaced by a single experimental stimulus, which appeared in the center of the display. Meaningful objects within each display appeared at the same location on the screen for non-competing as for competing configurations. Thus, the central contour

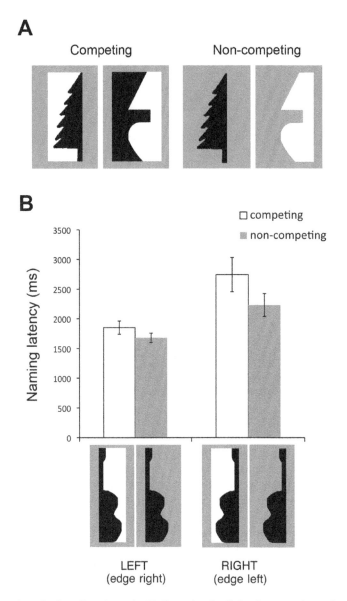

Figure 5. Example stimuli and results from Experiment 3. (A) Example stimuli for the competing and non-competing configurations. Objects were presented on either the left or right of the display, and the shade of the object was also manipulated. (B) Mean naming latency for correctly identified objects as a function of competition and side of display. Error bars represent 1 standard error of the mean. Note that RH failed to identify 24 of the 192 figures, 15 from displays with a right-sided shape (edge on the left) and 9 from displays with a left-sided shape (edge on the right).

defining the shape of the figure was located in precisely the same position for both competing and non-competing displays.

At the beginning of the experiment, RH was again shown the set of whole-object stimuli from Experiments 1 and 2 (now displayed on the computer screen) and asked to identify them, which he was able to do without error. RH was then shown each of the 192 experimental stimuli in turn, and cued to the shade of the object to be identified (black or white) before the onset of each new stimulus. RH's vocal response latency was recorded, and no feedback was given on accuracy.

Results and discussion

For the purpose of analysis we considered naming latencies only for those objects that were correctly identified across all conditions (i.e., regardless of side presentation or configuration). This constituted 16 of the 24 different objects. Figure 5 shows the mean naming latencies for competing and non-competing versions of shapes on the left and right sides of the central edge. Overall, RH was significantly slower to identify objects on the right vs. the left within both the competing and the non-competing conditions (Wilcoxon test, $p < .01$ for both comparisons), replicating the results from Experiments 1 and 2. Crucially, however, RH was also significantly slower to identify objects on the right when these were presented in competing versus non-competing displays (Wilcoxon test, $p < .05$). Thus, when a right-sided object's left edge was shared with a competing background region of opposite polarity (as in Experiments 1 and 2), RH took longer to identify the object than when it appeared on the same side, but in isolation against a surrounding gray background. In contrast, there was no significant difference between naming latencies for competing vs. non-competing displays when the object appeared on the left side of the display (Wilcoxon test, $p > .10$). The same trend was also apparent in RH's errors in identification. Of the 192 stimuli presented, RH failed to identify 15 from displays containing a shape on the right (5 competing, 10 non-competing) and 9 from displays with a shape on the left (4 competing, 5 non-competing).

As we predicted, RH's difficulty identifying shapes to the right, in which the critical contour defines the object's left edge, is exacerbated by a competing background region on the left side. It is noteworthy, however, that even for non-competing displays, RH named objects on the right significantly more slowly than shapes on the left, implying that removal of the competing background shape did not completely overcome his lateralized object recognition impairment.

GENERAL DISCUSSION

In the present case study, we explored the interaction between attention, figure–ground segmentation and object recognition in a neglect patient. When presented with a single contour dividing a figure–ground display into two equal regions, patient RH tended to perceive the leftward region as the figure even when the adjacent region on the right represented a meaningful and familiar object. This bias towards assigning figural status to shapes on the left of a contour supports previous findings of biased edge-assignment in neglect patients (Driver and Baylis, 1996; Marshall and Halligan, 1994; Peterson et al., 1998). Peterson et al. found that presenting identifiable figures can modulate this bias, from

which they inferred that object recognition processes precede figural parsing. Interestingly, we found that edge assignment and parsing the figure from ground does not appear to automatically support object recognition—confirming a trend reported by Peterson et al. (1998). Experiment 2 indicated that even when the correct figural assignment is cued, the patient continues to have difficulty in consciously identifying the parsed figure when this falls to the right of the critical dividing edge, so that this edge is assigned to the left of the cued figure. Thus, although object familiarity can provide implicit cues to figure–ground segmentation (Peterson et al., 1998), overt recognition of a cued figure can fail in neglect, when the critical dividing edge is to the left of that figure. This may represent intact ventral stream object processing for such objects interacting with dorsal stream spatial representations in neglect (see Driver and Vuilleumier, 2001).

In Experiment 3, we examined whether RH's impairment in recognising objects on the right of the display could reflect an intrusive, automatic edge assignment to the leftward region in the presence of a perceptually competing background shape. In support of this, we found a significant interaction between competition and the side of the display on which the object appeared, such that removing the competing region on the left of the display significantly increased RH's speed of recognizing a real shape on the right. Importantly, omitting the competing "background" region reduced, but did not eliminate, RH's bias in figural assignment and object recognition. This is probably because it is impossible to present a figure without an abutting negative space that could theoretically compete for the status of figure. Only the salience of this region can be modulated, and we demonstrated that reducing the salience of the competing region also reduced the bias in overt object recognition in neglect. Reducing the salience of the competing shape may have had a "bottom-up" effect on recognition by resolving the ambiguity in edge-assignment at the level of scene segmentation (Peterson and Skow, 2008). Alternatively, it may have influenced the way attention was deployed to the two regions and biased edge-assignment in a "top-down" manner, as has been observed in healthy participants' performance (Baylis and Driver, 1996; Vecera et al., 2004). Regardless, the present results demonstrate that neglect affects not only edge-assignment when segmenting figure from background, but also the conscious recognition of parsed figures. Interestingly, the relative perceptual salience of figure and ground can modulate the identification bias observed in neglect.

REFERENCES

Baylis, G. C., & Driver, J. (1996). One-sided edge assignment in vision: 1. Figure-ground segmentation and attention to objects. *Current Directions in Psychological Science, 4*, 140–146.

Behrmann, M., & Moscovich, M. (1994). Object-centred neglect in patients with unilateral neglect: Effect of left–right coordinates of objects. *Journal of Cognitive Neuroscience, 6*, 151–155.

Bisiach, E., & Luzzati, C. (1978). Unilateral neglect of representational space. *Cortex, 14*, 129–133.

Corbetta, M., & Shulman, G. L. (2002). Control of goal-directed and stimulus-driven attention in the brain. *Nature Reviews Neuroscience, 3*, 201–215.

Driver, J., & Baylis, G. C. (1996). One-sided edge assignment in vision: 2. Part decomposition, shape description, and attention to objects. *Current Directions in Psychological Science, 4*, 201–206.

Driver, J., & Halligan, P. W. (1991). Can visual neglect operate in object-centred coordinates? An affirmative single-case study. *Cognitive Neuropsychology, 8*, 475–496.

Driver, J., & Mattingley, J. B. (1998). Parietal neglect and visual awareness. *Nature Neuroscience, 1*, 17–22.

Driver, J., & Vuilleumier, P. (2001). Perceptual awareness and its loss in unilateral neglect and extinction. *Cognition, 79*, 39–88.

Driver, J., Baylis, G. C., & Rafal, R. D. (1992). Preserved figure–ground segregation and symmetry perception in visual neglect. *Nature, 360*, 73–75.

Gainotti, G., & Tiacci, C. (1970). Patterns of drawing disability in right and left hemispheric patients. *Neuropsychologia, 8*, 379–384.

Gibson, B., & Peterson, M. A. (1994). Does orientation-independent object recognition precede orientation-dependent recognition? Evidence from a cuing paradigm. *Journal of Experimental Psychology: Human Perception & Performance, 20*, 299–316.

Halligan, P. W., Fink, G. R., Marshall, J. C., & Vallar, G. (2003). Spatial cognition: Evidence from visual neglect. *Trends in Cognitive Sciences, 7*, 125–133.

Karnath, H.-O., Niemeier, M., & Dichgans, J. (1998). Space exploration in neglect. *Brain, 121*, 2357–2367.

Kim, C.-Y., & Blake, R. (2005). Psychophysical magic: Rendering the visible 'invisible'. *Trends in Cognitive Sciences, 9*, 381–388.

Kimchi, R., & Peterson, M. A. (2008). Figure–ground segmentation can occur without attention. *Psychological Science, 19*, 660–668.

Marshall, J. C., & Halligan, P. W. (1994). The Yin and Yang of visuo-spatial neglect: A case study. *Neuropsychologia 32*, 1037–1057.

Mesulam, M.-M. (1999). Spatial attention and neglect: Parietal, frontal and cingulate contributions to the mental representation and attentional targeting of salient extrapersonal events. *Philosophical Transactions of the Royal Society of London B: Biological Sciences, 354*, 1325–1346.

Peterson, M. A., & Gibson, B. S. (1994). Must figure–ground organization precede object recognition? An assumption in peril. *Psychological Science, 5*, 253–259.

Peterson, M. A., & Salvagio, E. (2008). Inhibitory competition in figure–ground perception: Context and convexity. *Journal of Vision, 8*, 1–13.

Peterson, M. A., & Skow, E. (2008). Suppression of shape properties on the ground side of an edge: Evidence for a competitive model of figure assignment. *Journal of Experimental Psychology: Human Perception & Performance, 34*, 251–267.

Peterson, M. A., Gerhardstein, P. C., Mennemeier, M. S., & Rapcsak, S. Z. (1998). Object-centred attentional biases and object recognition contributions to scene segmentation in left- and right-hemisphere damaged patients. *Psychobiology, 26*, 357–370.

Sejnowski, T. J., & Hinton, G. E. (1987). Separating figure from ground with a Boltzmann machine. In M. Arbib & A. Hanson (Eds.), *Vision, brain, and cooperative computation* (pp. 703–724). Cambridge, MA: MIT Press.

Snodgrass, J. G., & Vanderwart, M. (1980). A standardised set of 260 pictures. Norms for name agreement, image agreement, familiarity, and visual complexity. *Journal of Experimental Psychology: Human Learning and Memory, 6*, 174–215.

Stone, S. P., Halligan, P. W., & Greenwood, P. M. (1993). The incidence of neglect phenomena and related disorders in patients with an acute right or left hemisphere stroke. *Age and Ageing, 22*, 46–52.

Vecera, S., Flevaris, A., & Filapek, J. (2004). Exogenous spatial attention influences figure–ground assignment. *Psychological Science, 15*, 20–26.

Zipser, K., Lamme, V. A. F., & Schiller, P. H. (1996). Contextual modulation in primary visual cortex. *Journal of Neuroscience, 16*, 7376–7389.

COGNITIVE NEUROSCIENCE, 2010, 1 (3), 165–175

Theta-burst transcranial magnetic stimulation to the prefrontal cortex impairs metacognitive visual awareness

Elisabeth Rounis[1], Brian Maniscalco[2], John C. Rothwell[1],
Richard E. Passingham[1,3], and Hakwan Lau[1,2,3]

[1]University College London, London, UK
[2]Columbia University in the City of New York, New York, USA
[3]University of Oxford, Oxford, UK

We used a recently developed protocol of transcranial magnetic stimulation (TMS), theta-burst stimulation, to bilaterally depress activity in the dorsolateral prefrontal cortex as subjects performed a visual discrimination task. We found that TMS impaired subjects' ability to discriminate between correct and incorrect stimulus judgments. Specifically, after TMS subjects reported lower visibility levels for correctly identified stimuli, as if they were less fully aware of the quality of their visual information processing. A signal detection theory analysis confirmed that the results reflect a change in metacognitive sensitivity, not just response bias. The effect was specific to metacognition; TMS did not change stimulus discrimination performance, ruling out alternative explanations such as TMS impairing visual attention. Together these results suggest that activations in the prefrontal cortex in brain imaging experiments on visual awareness are not epiphenomena, but rather may reflect a critical metacognitive process.

Keywords: Transcranial magnetic stimulation; Prefrontal cortex; Visual awareness; metacognitive; stimulus judgments.

INTRODUCTION

Current theories suggest that the prefrontal cortex plays an important role in visual awareness (e.g. Crick & Koch, 1995; Dehaene, Sergent, & Changeux, 2003).

This hypothesis has been supported by a number of brain imaging studies (Marois, Yi, & Chun, 2004; Rees, Kreiman, & Koch, 2003; Sahraie et al., 1997). However, critics have claimed that there has been a lack of neuropsychological demonstration of the

Correspondence should be addressed to: Brian Maniscalco, Columbia University, Department of Psychology, 1190 Amsterdam Avenue, MC 5501, New York, NY 10027, USA.
E-mail: brian@psych.columbia.edu
This work is supported by Sun-Chan and internal funding from Columbia University (to HL), and by the Wellcome Trust (to REP).

www.psypress.com/cognitiveneuroscience
DOI: 10.1080/17588921003632529

essential role of the prefrontal cortex in visual awareness: Damage to the prefrontal cortex does not seem to lead to cortical blindness (Pollen, 1995). Here we attempt to clarify this issue by showing that bilateral TMS to the prefrontal cortex does have an effect on visual awareness, in particular the metacognitive sensitivity with which it discriminates between effective and ineffective stimulus processing. While we agree with critics that disruption of prefrontal activity may have little or no effect on primary visual awareness, i.e., the ability to represent visual targets, higher monitoring aspects of awareness may critically depend on prefrontal activity.

Recent studies have shown that visual awareness can be assessed by metacognitive procedures (Kolb & Braun, 1995; Persaud, McLeod, & Cowey, 2007). Typically, when awareness is lacking, one tends to place low or inappropriate subjective ratings (Weiskrantz, 1997). Such metacognitive approaches have also been used in other studies of visual awareness (Galvin, Podd, Drga, & Whitmore, 2003; Kolb & Braun, 1995; Lau & Passingham, 2006; Szczepanowski & Pessoa, 2007) and implicit learning (Dienes & Perner, 1999; Persaud et al., 2007). In general, one can assess metacognitive sensitivity by measuring how well subjective ratings (e.g., of confidence or visibility) distinguish

between correct and incorrect judgments (e.g., about the identity of a presented stimulus). High levels of metacognitive sensitivity imply that subjects are introspectively aware of the effectiveness of their internal information processing (Kolb & Braun, 1995; Galvin et al., 2003). As there is not yet widespread agreement on the ideal measure of metacognitive sensitivity, in the present study we use two separate approaches—a correlation approach and a signal detection theory approach—and demonstrate converging interpretations of the data.

We required volunteers to perform a two-alternative forced-choice visual task, identifying the spatial arrangement of two visual stimuli (a square and a diamond, Figure 1A). At the same time, they also rated the subjective visibility of the stimuli ("clear" or "unclear"). Subjects performed these tasks before and after transcranial magnetic stimulation (TMS), which was aimed at the dorsolateral prefrontal cortex (DLPFC, Figure 1B). We applied to this region theta-burst stimulation (TBS), a recently developed protocol that is known to effectively depress cortical excitability by mimicking the action of long-term potentiation and long-term depression in cortical tissues (Huang et al., 2005). One advantage of this technique is that the effect of 20 s of stimulation is known to last for up to

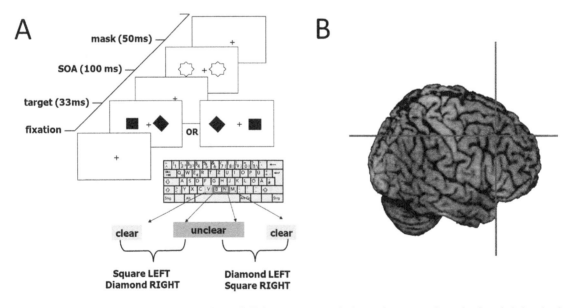

Figure 1. Experimental design. (A) Visual task and stimuli. Volunteers were required to perform a two-alternative forced-choice visual task, identifying the spatial arrangement of two visual stimuli (square on the left and diamond on the right, or the other way round). They rated the subjective visibility ("clear" or "unclear") at the same time. So in every trial subjects had four options as to which key to press in order to respond. (B) Site of stimulation. The dorsolateral prefrontal cortex (DLPFC) was the targeted site of stimulation, and was chosen because neural activity from this area has been shown to reflect a difference in the subjective ratings of visibility even when performance in a forced-choice visual task was matched (Lau & Passingham, 2006). The image showing the site of stimulation is based on magnetic resonance brain scans of 6 of the 20 subjects in this study. The scans were collected after completion of the TMS experiments. Right and left DLPFC coordinates were [37 26 50] and [–41 18 52], with standard deviations [4.6 5.6 5.3] and [4.3 5.1 3.8] respectively.

20 min, which means we had the opportunity to depress both sides of the DLPFC by stimulating them sequentially. We opted for bilateral stimulation as this has been suggested to be critical: Sahraie et al. (1997) have suggested that one reason visual defects do not seem to frequently follow prefrontal lesions may be that such lesions have to be large and bilateral. Using this sequential method to depress the DLPFC bilaterally, we found that the metacognitive sensitivity of reported visual awareness was reduced after TMS.

METHOD

Subjects

Twenty healthy volunteers (eight women, mean age 25.6 ± 6.1), with normal or corrected-to-normal vision and no history of neurological disorders or head injury were recruited from the database of volunteers at the Functional Imaging Laboratory, Institute of Neurology, University College London. Written informed consent was obtained from all participants. The study was approved by the joint ethics committee for the National Hospital for Neurology and Neurosurgery (UCLH NHS Trust) and the Institute of Neurology (UCL).

Experimental design

Subjects were asked to perform a two-alternative forced-choice task (Figure 1A). Testing was performed in a darkened room. Stimuli were presented against the white background of a CRT monitor refreshing at 120 Hz. The monitor was placed 40 cm away from the subjects' eyes. On each trial, a diamond and a square were presented on either side of a central crosshair for 33 ms. The stimuli had sides measuring $0.8°$ of visual angle and were centered $1°$ to the left and right of the central crosshair. 100 ms after stimulus onset, a metacontrast mask was displayed for 50 ms in order to enhance task difficulty. The two possibilities for the sequence of stimuli (square on the left and diamond on the right, and vice versa) were presented with equal probability in a pseudorandom order. The subjects' task was to identify which stimulus sequence had just been presented, square left/diamond right or vice versa. At the same time, subjects gave subjective ratings of stimulus visibility ("clear" or "unclear"). Subjects were instructed to make the visibility judgment in a relative manner, to distinguish between stimuli that were relatively

more or less visible. Since stimulus contrast was adjusted so as to yield threshold performance on the stimulus classification task, stimuli used in this experiment were somewhat difficult to see. Nonetheless, subjects were instructed to judge stimulus visibility on each trial relative to the context of stimuli used in this experiment. For instance, a subject might judge that the stimulus on a certain trial was more readily visible than the majority of stimuli seen in the experiment up to that point, even if its visibility was poor by everyday standards. Subjects were encouraged to judge such stimuli as exhibiting "high clarity," i.e., having relatively high clarity compared to other stimuli observed in the experimental context.

Subjects attended two separate testing sessions, both preceded by a demonstration and a practice phase of 100 trials intended to familiarize the subjects with the task and to allow them to reach a stable level of performance. Performance level was controlled to be at approximately 75% correct throughout the experiment by titrating the contrast of the stimuli, using a standard up–down transformed-response staircasing procedure (Macmillan & Creelman, 2005). Each trial was randomly designated as belonging to staircase A or staircase B. For staircase A, contrast on the current trial was increased if the subject responded incorrectly on the previous "A" trial, whereas contrast on the current trial was decreased if the subject responded correctly on the two previous "A" trials. Staircase B worked in a similar manner, except it required three consecutive correct responses on "B" trials in order to reduce contrast. After practice, subjects underwent an initial ("pre") block of 300 trials to measure forced-choice task performance and subjective ratings of visibility. On average this took 10.9 min, excluding brief breaks after every 100 trials. After completing this block, two real (or sham) continuous TBS (cTBS) conditioning stimulations, one to the left and one to the right, were delivered to the dorsolateral prefrontal area. The two stimulations were separated by a 1-min intertrain interval. Following real (or sham) stimulation, subjects did another ("post") block of 300 trials. On average this took 10.4 min, excluding brief breaks after every 100 trials. Session order by type of cTBS (real vs. sham) was counterbalanced across subjects.

Theta-burst stimulation

In each TBS session, 600 biphasic stimuli, at a stimulation intensity of 80% of active motor threshold (AMT) for the right first dorsal interosseous (FDI) hand muscle, were given over the left and right

DLPFC area using a Magstim Super Rapid stimulator (Whitland, UK) connected to four booster modules. The conditioning cTBS stimuli were delivered in two separate 20-s trains of 300 cTBS pulses, one for the left and one for the right, separated by an intertrain interval of 1 min. A similar bilateral procedure has been used in a recent clinical study (Artfeller, Vonthein, Plontke, & Plewnia, 2009).

A standard figure-of-eight-shaped coil (Double 70 mm Coil Type P/N 9925; Magstim) was used for both real and sham cTBS. Real cTBS was delivered with the coil placed tangentially to the scalp with the handle pointing posteriorly. In sham cTBS sessions, the coil was placed perpendicularly to the scalp, an ineffective position for the delivery of conditioning pulses, which provided comparable acoustic stimuli to the real cTBS condition. The coil was positioned with the handle at 45° to the sagittal plane. The current flow in the initial rising phase of the biphasic pulse in the biphasic pulse induced a posterior-to-anterior current flow in the underlying cortex.

The basic TBS pattern was a burst containing three pulses of 50 Hz magnetic stimulation given in 200 ms intervals (i.e. at 5 Hz). In the continuous theta burst stimulation paradigm (cTBS), a 20 s train of uninterrupted TBS is given (300 pulses or 100 bursts). Physiological studies have shown that this produces a decrease in corticospinal excitability which lasts for about 20 min (Huang et al., 2005), when applied to the primary motor cortex, M1. This new rTMS paradigm has been developed recently in our laboratory and has the advantage of being a rapid and efficient method of conditioning, which has effects on corticospinal excitability that have been shown to involve similar mechanisms to long-term potentiation/depression (LTP/LTD) with NMDA dependence (Huang et al., 2007), as well as effects on behavior and learning (Huang et al., 2005; Talelli, Greenwood, & Rothwell, 2007).

The site of cTBS stimulation was located 5 cm anterior to the "motor hot spot" on a line parallel to the midsagittal line. This DLPFC location has been used in previous studies and can be shown consistently on structural scans (Mottaghy, Gangitano, Sparing, Krause, & Pascual-Leone, 2002; Rounis et al., 2006; Figure 1B). The position of the motor hot spot was defined functionally as the point of maximum evoked motor response in the slightly contracted right FDI. The active motor threshold was defined as the lowest stimulus intensity that elicited at least five twitches in 10 consecutive stimuli given over the motor hot spot, while the subject was maintaining a voluntary contraction of about 20% of maximum using visual feedback.

The use of such low subthreshold intensity (80% AMT) had the advantage of decreased spread of stimulation away from the targeted site, thus keeping the area that was stimulated with the conditioning pulses more focal (Pascual-Leone, Valls-Solé, Wassermann, & Hallett, 1994; Münchau, Bloem, Irlbacher, Trimble, & Rothwell, 2002). Also, a previous study on the prefrontal cortex that applied intensity above motor threshold reported unpleasant vagal reactions in subjects (Grossheinrich et al., 2009). However, even at that higher intensity there was no adverse effects on mood, seizure or epileptiform observed in the recorded electroencephalogram. This suggests that our stimulation at this lower intensity should be safe to our subjects.

Data analysis

Metacognitive sensitivity (i.e. the efficacy with which visibility ratings distinguish between correct and incorrect responses) was assessed using two separate methods. The first method followed previous studies (e.g., Kolb & Braun, 1995; Kornell, Son, & Terrace, 2007) in using the correlation between accuracy and subjective rating as a measure of metacognitive sensitivity. We used the correlation coefficient phi, which quantifies the degree of correlation between two binary variables, to calculate the correlation between task accuracy (correct/incorrect) and stimulus visibility (clear/unclear). Phi is equivalent to Pearson's r computed for two binary variables, and like r it ranges from –1 (perfect negative correlation) to +1 (perfect positive correlation). We calculated phi for the 300 trials pre and post real and sham TMS for each subject. We predicted that TMS would hinder metacognitive sensitivity, and thus that there would be a TMS (real/sham) × time (pre/post) interaction.

We also performed a signal detection theoretic (SDT) analysis to estimate metacognitive sensitivity. On this analysis, we estimate a value called meta-d', which is the amount of signal available for one's metacognitive disposal (i.e. available for doing the confidence/visibility rating task). This measure is in the same scale as d', i.e., the signal that is available for the primary stimulus classification task, so the two can be directly compared. If meta-$d' < d'$, it means that some signal that is available for the primary stimulus classification task is lost for metacognition, which means that metacognitive sensitivity is not perfect.

The need to perform an SDT analysis is due to the fact that phi can be shown to generate non-regular receiver operating characteristic (ROC) curves, which in turn implies an underlying threshold model of detection (Swets, 1986). The ROC profile and threshold model of phi are not in good agreement with the standard SDT model (Macmillan & Creelman, 2005), nor with a recent treatment of the "type 2" signal detection model that characterizes metacognitive performance within an SDT framework (Galvin et al., 2003). (In the following, we use the terms "metacognitive" and "type 2" interchangeably.) The consequence of this is that phi may confound sensitivity and response bias, rather than being a pure measure of sensitivity. Thus, we also performed a type 2 SDT analysis of the data.

The fundamental idea in SDT is to use observed hit rate (HR) and false alarm rate (FAR) data to estimate an observer's sensitivity (d') and response bias (c) (Macmillan & Creelman, 2005; Figure 2A). Likewise, the fundamental idea of a type 2 SDT analysis is to use type 2 HR (i.e. the frequency with which correct responses are endorsed with high confidence) and type 2 FAR (i.e. the frequency with which incorrect responses are endorsed with high confidence) to estimate the sensitivity and response bias an observer's confidence ratings (or, analogously, visibility ratings) exhibit in classifying first-order stimulus judgments as correct or incorrect. However, the underlying formalisms for the type 2 SDT model are quite complex (Galvin et al., 2003), and there is as yet no widespread agreement on how to perform a proper type 2 SDT analysis. One could estimate the type 2 ROC curve, and measure type 2 sensitivity by the area under the curve. However, this measure would depend on type 1 d' (Galvin et al., 2003), and any observed change in type 2 sensitivity could therefore be confounded. We therefore developed a measure called meta-d', i.e. the signal that is available for the type 2 task, and compared it against type 1 d'. Below is how we calculate this meta-d' measure.

We characterize type 2 sensitivity by capitalizing on the observation that, on the classic SDT model, the sensitivity component of type 2 HR and FAR is already determined by the sensitivity for stimulus discrimination, d', in conjunction with the criterion for stimulus classification response, c (Galvin et al., 2003). That is, once d' and c for the classic SDT model are fixed, a type 2 ROC curve defining the tradeoff between type 2 HR and FAR is already implied for each type of stimulus response (Figure 2B). Thus, we characterize type 2 sensitivity as the d' an ideal SDT observer with a fixed stimulus classification criterion c would require in order to produce the observed type 2 HR and FAR data (Figure 2C). This measure (call it meta-d') can then be compared to the observer's actual d' in order to assess how well the observer's observed type 2 sensitivity compares to the theoretically ideal type 2 sensitivity, according to the classic SDT model. For an ideal SDT observer, meta-$d' = d'$; for suboptimal metacognitive sensitivity, meta-$d' < d'$; and for an observer whose confidence ratings are not diagnostic of judgment accuracy at all, meta-$d' = 0$. We predicted that the difference meta-$d' - d'$ would decrease more following real TMS than sham TMS, indicating that TMS hinders subjects' metacognitive sensitivity, relative to the theoretical ideal.

We estimated meta-d' for each subject, in each TMS × time condition, as follows. First, we estimated the SDT parameters c' (the stimulus classification criterion measured relative to d') and s (the ratio of standard deviations of internal evidence for the two stimulus classes) (Macmillan & Creelman, 2005; Figure 2A). Holding c' and s constant, we estimated values for meta-d' and the type 2 criterion $c_{\text{conf | response="square left/diamond right"}}$ that would minimize the sum of squared errors between observed and modeled type 2 HR and FAR for all trials where the subject classified the stimulus as "square left/diamond right." We then estimated values for meta-d' and the type 2 criterion $c_{\text{conf | response="diamond left/square right"}}$ that would minimize the sum of squared errors between observed and modeled type 2 HR and FAR for all trials where the subject classified the stimulus as "diamond left/square right." (Compare to Figure 2, where, for example, "square left/diamond right" corresponds to "S1" and "diamond left/square right" corresponds to "S2"). Thus, we generated two estimates of meta-d', corresponding to the subject's type 2 HR and FAR conditional on each stimulus classification type. The two estimates were combined via a weighted average, where the weight of each meta-d' estimate was determined by the number of trials used to estimate it. The mean SSE corresponding to each meta-d' estimate was 9.1×10^{-5}, indicating that this approach provided an excellent fit to the observed type 2 HR and FAR data. Minimization of SSE was achieved using the Optimization Toolbox in MATLAB® (MathWorks, Natick, MA).

Because we are testing a directional hypothesis in a 2 × 2 factorial design (i.e., metacognitive sensitivity is reduced following real TMS more than following sham TMS), we report halved p-values for the TMS × time interaction on phi and meta-$d' - d'$.

RESULTS

In the following we present ANOVA analyses with within-subject factors of TMS (real/sham) and time (pre/post) for several independent variables of interest such as accuracy and response time for correct

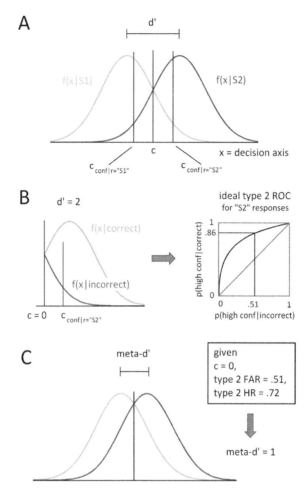

Figure 2. SDT analysis of type 2 (metacognitive) performance. The basic idea of this analysis is to compute meta-d', a measure of the signal that is available for one's metacognitive disposal (i.e., available for making subjective ratings). This measure is in the same scale as d', i.e., the signal available for the primary forced-choice task, such that the two can be directly compared. If meta-$d' < d$, it means that some signal that is available for the primary forced-choice task is lost in the rating, i.e. the subject is not metacognitively perfect. (A) The classic SDT model. The observer must discriminate between stimulus classes S1 and S2. Each stimulus presentation generates a value on an internal decision axis, corresponding to the evidence in favor of S1 or S2. Evidence generated by each stimulus class is normally distributed across the decision axis, and the normalized distance between these distributions (d') measures how well the observer can discriminate S1 from S2. The observer sets a decision criterion c, such that all signals exceeding c are labeled "S2" and all those failing to exceed c "S1." The observer also sets criteria $c_{conf \mid r="S1"}$ and $c_{conf \mid r="S2"}$ to determine confidence ratings (higher ratings for signals farther from c). For expositional ease in this hypothetical example, we set $d' = 2$, $c = 0$, and s (the ratio of standard deviations for the two distributions) = 1; in the actual analysis, d', c, and s are estimated from data. (B) Type 2 sensitivity from d' and c. Consider only trials where the observer responds "S2." Then the S2 distribution corresponds to the distribution of evidence for correct responses, and the S1 distribution corresponds to the distribution of evidence for incorrect responses. All trials surpassing $c_{conf \mid r="S2"}$ are endorsed with high confidence. Sweeping the $c_{conf \mid r="S2"}$ criterion across the decision axis generates different values for type 2 false alarm rate, $p($ high confidence | incorrect), and type 2 hit rate, $p($ high confidence | correct), and thus generates a type 2 ROC curve. (Similar considerations hold for "S1" responses.) Thus, d' and c are jointly sufficient to determine type 2 sensitivity for each response type, according to the standard SDT model. (C) Characterizing type 2 sensitivity. The analysis from (A) and (B) can be reversed to characterize metacognitive sensitivity. Suppose that the observer produced type 2 FAR = .51, type 2 HR = .72. We then ask: "What d' would an SDT-optimal observer with stimulus classification criterion c require in order to produce the observed type 2 FAR and type 2 HR?" The answer is meta-d', a characterization of the sensitivity with which confidence ratings discriminate between correct and incorrect judgments. In the example in the figure, meta-$d' = 1$ even though observed $d' = 2$, indicating suboptimal metacognitive sensitivity.

trials. None of these analyses exhibited a main effect of TMS condition (F values < 1.7), indicating that the real and sham TMS sessions were comparable on baseline task performance.

Stimulus contrast was adjusted online in order to control classification accuracy; thus, as expected, frequency of correct responses did not vary as a function of time, $F(1, 19) = 2.45$, $MSE = 0.001$) or the TMS × time interaction, $F(1, 19) = 0.002$, $MSE = 0.001$ (Figure 3A). A more insightful measure of stimulus classification performance is the mean contrast required to keep classification accuracy

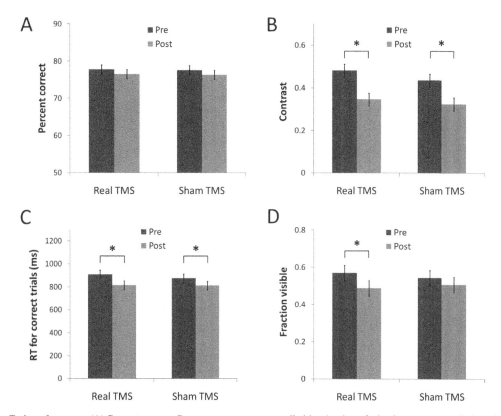

Figure 3. Task performance. (A) Percent correct. Percent correct was controlled by titration of stimulus contrast, such that stimulus judgments were about 75% correct throughout the experiment (see "Method"). Therefore the lack of any significant effects on these values is trivial. (B) Mean stimulus contrast. Stimulus contrast was determined online by the computer program (see "Method"), such that if subjects performed better than 75% correct, the contrast was reduced, and if subjects performed worse than 75% correct, the contrast was increased. There was a main effect of time on contrast ($p < .001$), indicating a perceptual learning effect; had the computer not been programmed to adjust task difficulty online, subjects would have shown improved accuracy over time. However, perceptual learning was not affected by TMS (TMS × time interaction, $F = 0.73$). (C) Reaction time for correct responses. Perceptual learning was also evident in reaction time data. Subjects were quicker to make correct responses in the second half of the experiment (main effect of time, $p = .016$). However, again, this learning effect was not modulated by TMS (TMS × time interaction, $F = 0.79$). (D) Mean visibility ratings. Visibility ratings decreased over time ($p = .005$), but the TMS × time interaction on visibility was not significant ($p = .4$). See the discussion for caveats about the visibility rating analysis. "Real pre": performance level before real TMS. "Real post": after real TMS. "Sham pre": before sham TMS. "Sham post": after sham TMS. *$p < .05$. Error bars represent 1 *SEM*.

constant. The stimulus contrast generated by the performance staircasing algorithm reduced over time, $F(1, 19) = 88.06$, $MSE = 0.003$, $p < .001$, suggesting a perceptual learning effect: Over time, subjects required a lower level of contrast in order to maintain the same level of response accuracy. However, the TMS × time interaction was not significant, $F(1, 19) = 0.73$, $MSE = 0.004$, indicating that the TMS treatment had no effect on stimulus classification performance (Figure 3B). Likewise, reaction time for correct trials improved over time, $F(1, 19) = 7.04$, $MSE = 17863$, $p = .016$, but was not sensitive to TMS, $F(1, 19) = 0.80$, $MSE = 6064$ (Figure 3C).

Similarly, mean visibility ratings decreased over time, $F(1, 19) = 9.92$, $MSE = 0.008$, $p = .005$, but independently of the TMS manipulation, $F(1, 19) = 1.2$,

$MSE = 0.007$ (Figure 3D). We address this null finding more fully in the discussion.

As hypothesized, TMS significantly impaired metacognitive sensitivity. A TMS × time interaction was evident for the correlation between accuracy and visibility, phi, $F(1, 19) = 3.64$, $MSE = 0.002$, $p = .036$ (Figure 4A). Investigation of this interaction revealed that phi was lowered following real TMS, one-tailed paired t-test, $t(19) = 4.13$, $p < .001$, but not sham TMS, $t(19) = 0.77$.

The bias-free SDT measure of metacognitive sensitivity, meta-$d' - d'$, also exhibited a TMS × time interaction effect, $F(1, 19) = 5.51$, $MSE = 0.125$, $p = .015$ (Figure 4B). The difference between observed and ideal type 2 sensitivity decreased following real TMS, one-tailed paired t-test, $t(19) = -3.1$, $p = .006$, but not sham, $t(19) = 0.37$. Metacognitive sensitivity

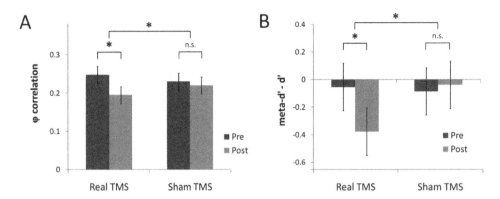

Figure 4. Effect of TMS on metacognitive sensitivity. (A) Correlation coefficient, phi. TMS significantly reduced phi, the correlation between stimulus classification accuracy and stimulus visibility. The effect of TMS is evident in a significant TMS × time interaction, $p = .036$; phi was lower following real TMS ($p < .001$) but not sham TMS ($p = .5$). (B) Divergence from optimal metacognitive sensitivity, meta-$d' - d'$. A signal detection theory analysis revealed that subjects' metacognitive sensitivity, relative to the optimal level of metacogntive sensitivity determined by their task performance (see Figure 2 and "Method"), was significantly impaired by TMS (interaction $p = .015$). Metacognitive sensitivity was lower following real TMS ($p = .006$) but not sham TMS ($p = .7$). Subjects exhibited significantly suboptimal metacognitive sensitivity following real TMS, i.e., meta-$d' - d' < 0$ ($p = .004$) but not in any other experimental condition (p values > .3). "Real pre": metacognitive performance before real TMS. "Real post": after real TMS. "Sham pre": before sham TMS. "Sham post": after sham TMS. *$p < .05$; n.s., not significant. Error bars represent 1 *SEM*.

was significantly suboptimal following real TMS, i.e. meta-$d' < d'$, one-tailed t-test, $t(19) = 2.93$, $p = .004$, but not in any other TMS × time condition (t values < 1).

There are several ways in which TMS could have impaired metacognitive sensitivity. One possibility is that TMS reduced visibility for correct trials, which would amount to a kind of relative blindsight (Lau & Passingham, 2006). Alternatively, TMS may

have increased visibility for incorrect trials, a kind of "hallucinatory" effect. A third possibility is that the reduction in metacognitive sensitivity was not specific to correct or incorrect trials. Thus, to better characterize the effect of TMS, we examined visibility ratings separately for correct and incorrect trials pre- and post-TMS (Figure 5A). We found a significant accuracy × time interaction, $F(1, 19) = 18.04$, $MSE = 0.002$, $p < .001$, driven by the fact that TMS

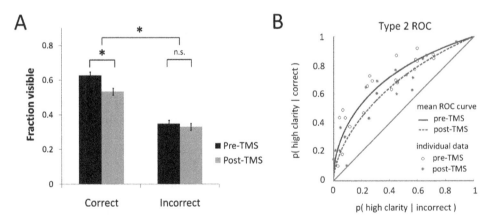

Figure 5. Nature of the TMS effect on metacognition. (A) Selective reduction of type 2 hit rate. Visibility ratings are displayed as a function of time (pre-/post-TMS) and accuracy (correct/incorrect) for the real TMS condition. TMS significantly reduced visibility for correct responses (two-tailed paired t-test, $p = .002$), but not for incorrect responses ($p = .5$). The time × accuracy interaction was significant, $p < .001$. These results suggest that TMS reduced metacognitive sensitivity (Figure 4) specifically by decreasing visibility ratings for correct responses (as opposed to increasing visibility ratings for incorrect responses). Thus, TMS induced a kind of relative blindsight, to the extent that TMS suppressed the reports of visibility for accurately processed stimuli. *$p < .005$; n.s., not significant. Error bars represent 1 *SEM*. (B) Type 2 ROC analysis. Individual data points indicate the type 2 hit rates and false alarm rates for every subject pre- and post-TMS. Type 2 ROC curves were estimated for each subject using estimates of meta-d', c', and s; the average of these ROC curves is plotted for the pre- and post-real TMS conditions. The distribution of individual data points and the fitted ROC curves indicate that TMS influenced metacognitive sensitivity rather than just response bias. Note that the ROC data is a reflection of meta-d', and thus is not as sensitive to the effect of TMS as the measure used in the analysis, meta-$d' - d'$ (Figure 4), since some variation in meta-d' is attributable merely to variation in d' (Figure 2).

reduced visibility for correct responses, two-tailed paired t-test, $t(19) = 3.54$, $p = .002$, but not incorrect responses, $t(19) = 0.72$. Thus, TMS impaired metacognitive sensitivity by selectively reducing the visibility of correctly classified stimuli.

DISCUSSION

Our results show that theta-burst TMS applied to bilateral DLPFC can reduce metacognitive sensitivity, i.e. the efficacy with which subjective visibility ratings distinguish between correct and incorrect stimulus judgments. This effect was driven specifically by a reduction in visibility for correct trials, rather than by a specific elevation of visibility for incorrect trials or by a nonspecific effect. In this sense, the direction of the effect is reminiscent of blindsight (Weiskrantz, 1997), where patients deny visual awareness even when they can perform visual discrimination tasks well above chance level. The effect of TMS was specific to metacognitive sensitivity; TMS did not disrupt stimulus classification performance, as measured by contrast level (Figure 3B) and reaction time for correct trials (Figure 3C).

We did not find a significant effect of TMS on averaged stimulus visibility itself. However, note that the effect of TMS is at least partially characterized by a change in visibility ratings, in that TMS reduced metacognitive sensitivity precisely by reducing visibility for correctly classified stimuli while leaving visibility for incorrectly classified stimuli unaffected (Figure 5A). Indeed, although the interaction was not significant, separate paired t-tests show a difference in visibility pre and post real TMS, two-tailed, $t(19) = 3.09$, $p = .002$, but no difference pre and post sham TMS, $t(19) = 1.47$. There are two reasons why the design of the current study may not have been ideal to statistically detect an effect of TMS on overall stimulus visibility. One is that stimulus visibility was affected by an experimental factor other than TMS, namely the contrast levels of the stimuli, which were adjusted online throughout the experiment in order to hold discrimination performance constant. Another reason is that subjects were instructed to use visibility ratings in a relative manner, in order to distinguish stimuli that were relatively more or less visible. The instruction to rate visibility in this relative way may have obscured the extent to which visibility ratings reflected absolute differences in stimulus visibility across experimental conditions. Nonetheless, these limitations are not important for the main focus of this study, which is the metacognitive sensitivity of visibility ratings.

One typical argument against studies of awareness is that the manipulation in question might have only changed subjects' criteria for producing subjective ratings, rather than changing awareness *per se*. A change in response criterion is not necessarily uninteresting, but, more importantly, this is not what we found. Our type 2 SDT analysis demonstrates that TMS reduced metacognitive sensitivity (i.e. the efficacy with which subjective visibility ratings discriminate between correct and incorrect judgments), rather than merely affecting metacognitive response bias (i.e. the overall propensity to give high visibility ratings). TMS reduced visibility for correct trials (type 2 HR) but not for incorrect trials (type 2 FAR) (Figure 5A), a pattern that cannot be attributed solely to changes in response bias. Likewise, our measure of type 2 sensitivity, meta-d' − d', is not sensitive to changes in type 2 response bias (Figure 2). We also demonstrate this point graphically in Figure 5B, which shows the type 2 ROC points for each subject, and mean fitted ROC curves, pre- and post-TMS. The distributions of type 2 ROC points and the fitted type 2 ROC curves differ, indicating lower type 2 sensitivity following TMS. If TMS only affected subjects' criteria for reporting high visibility, the type 2 ROC curves pre- and post-TMS should overlap (Macmillan & Creelman, 2005), contrary to our findings.

Because we only used visual stimuli, we cannot rule out the interesting possibility that the observed deficit in metacognitive sensitivity following TMS to DLPFC would apply to tasks in other modalities, e.g. auditory tasks. It may be that the observed reduction in metacognitive sensitivity is not specific to visual processes *per se*, but rather generalizes to other decision-making contexts involving confidence or perceptual clarity judgments. However, note that our analysis rules out the possibility that such a nonspecific effect could be carried by general influences on, for example, risk aversion or overall confidence level. Such differences would constitute differences in type 2 response bias, not type 2 sensitivity.

Our results extend previous work. Similarly to the present study, Del Cul, Dehaene, Reyes, Bravo, and Slachevsky (2009) showed that prefrontal lesions can affect subjective reports of visual experience more than visual task performance. Slachevsky et al. (2001, 2003) have shown that lesion to the prefrontal cortex can affect awareness in the monitoring of actions or sensory-motor readjustments. Other studies show that visual processing can be affected by lesion (Latto & Cowey, 1971) or TMS (Grosbras & Paus, 2003; Ruff et al., 2006) to the frontal eye field. Turatto, Sandrini, and Miniussi (2004) showed that TMS to the DLPFC can affect performance in change blindness. These studies show that, contrary to what critics have argued (Pollen, 1995), disruption of activity in the prefrontal cortex can in fact influence awareness and visual processing. What is new in the present study is that it specifically highlights the role of the prefrontal cortex in supporting the metacognitive sensitivity of visual awareness.

The prefrontal cortex is associated with many important cognitive functions, and therefore our interpretation is not that it is completely specific to the metacognitive sensitivity of visual awareness. It is likely that bilateral theta-burst TMS to the DLPFC would impair performance in other tasks where metacognitive visual awareness is not required. For instance, as mentioned above, we think it is likely that it may also impair metacognitive awareness for auditory signals. Instead of applying the same TMS treatment to unrelated control tasks and hoping to show a negative result in those situations, we show that TMS impaired a specific *process* involved in our task, namely metacognitive awareness, but not other processes involved in the same task. It is important to note that performance in the stimulus classification task was not influenced by TMS under the stimulation parameters currently used. Thus, it is unlikely that TMS affected metacognitive sensitivity by means of nonspecific disturbances such as reductions in visual attention or general arousal.

As in Del Cul et al. (2009), one limitation of the present study is that we did not show that a similar effect could not be obtained in a control anatomical site. The lack of such control conditions is unfortunate and largely constrained by logistics (e.g., we did not have ethical approval for every brain region for this relatively new TMS protocol, and the leading authors have since relocated). However, given that the TMS was applied offline (i.e., not during task), and that the effect did not change basic task performance, it is unlikely that the results we obtained were due to the general distraction due to TMS. It is likely that TMS applied to an unrelated region, such as the somatosensory area, would not lead to our metacognitive effect. However, it remains an open question whether TMS applied to parietal areas that are connected to DLPFC would lead to similar results.

In any case, our conclusion is not that the neural circuitry that supports metacognitive visual awareness is completely localized in the DLPFC. Rather, we conclude that disruption of activity in this area can impair the metacognitive sensitivity of visual awareness. The present results show that the prefrontal cortex is functionally relevant to visual awareness, in that manipulation of the former can affect the latter.

Further, the data clarify in what way the prefrontal cortex might contribute. Activity in the DLPFC may play a relatively unimportant role in representing the visual signal itself, but it may be essential for some form of internal uncertainty monitoring that allows observers to be able to distinguish when visual processing is effective and when it is not. It is this introspective and metacognitive aspect of visual awareness for which the prefrontal cortex may be critical.

REFERENCES

Arfeller, C., Vonthein, R., Plontke, S. K., & Plewnia, C. (2009). Efficacy and safety of bilateral continuous theta burst stimulation (cTBS) for the treatment of chronic tinnitus: Design of a three-armed randomized controlled trial. *Trials, 10*, 74.

Arnold, D. H., Law, P., & Wallis, T. S. A. (2008). Binocular switch suppression: A new method for persistently rendering the visible 'invisible'. *Vision Research, 48*(8), 994–1001.

Crick, F., & Koch, C. (1995). Are we aware of neural activity in primary visual cortex? *Nature, 375*(6527), 121–123.

Dehaene, S., Sergent, C., & Changeux, J. (2003). A neuronal network model linking subjective reports and objective physiological data during conscious perception. *Proceedings of the National Academy of Sciences of the United States of America, 100*(14), 8520–8525.

Del Cul, A., Dehaene, S., Reyes, P., Bravo, E., & Slachevsky, A. (2009). Causal role of prefrontal cortex in the threshold for access to consciousness. *Brain, 132*(9), 2531–2540.

Dienes, Z., & Perner, J. (1999). A theory of implicit and explicit knowledge. *Behavioral and Brain Sciences, 22*(5), 735–755; discussion, 755–808.

Galvin, S. J., Podd, J. V., Drga, V., & Whitmore, J. (2003). Type 2 tasks in the theory of signal detectability: Discrimination between correct and incorrect decisions. *Psychonomic Bulletin & Review, 10*(4), 843–876.

Grosbras, M., & Paus, T. (2003). Transcranial magnetic stimulation of the human frontal eye field facilitates visual awareness. *European Journal of Neuroscience, 18*(11), 3121–3126.

Grossheinrich, N., Rau, A., Pogarell, O., Hennig-Fast, K., Reinl, M., Karch, S., et al. (2009). Theta burst stimulation of the prefrontal cortex: Safety and impact on cognition, mood, and resting electroencephalogram. *Biological Psychiatry, 65*(9), 778–784.

Huang, Y., Edwards, M. J., Rounis, E., Bhatia, K. P., & Rothwell, J. C. (2005). Theta burst stimulation of the human motor cortex. *Neuron, 45*(2), 201–206.

Huang, Y., Rothwell, J. C., Edwards, M. J., & Chen, R. (2008). Effect of physiological activity on an NMDA-dependent form of cortical plasticity in human. *Cerebral Cortex, 18*(3), 563–570.

Kolb, F. C., & Braun, J. (1995). Blindsight in normal observers. *Nature, 377*(6547), 336–338.

Kornell, N., Son, L. K., & Terrace, H. S. (2007). Transfer of metacognitive skills and hint seeking in monkeys. *Psychological Science*, *18*(1), 64–71.

Latto, R., & Cowey, A. (1971). Visual field defects after frontal eye-field lesions in monkeys. *Brain Research*, *30*(1), 1–24.

Lau, H. C., & Passingham, R. E. (2006). Relative blindsight in normal observers and the neural correlate of visual consciousness. *Proceedings of the National Academy of Sciences of the United States of America*, *103*(49), 18763–18768.

Macmillan, N. A., & Creelman, C. D. (2005). *Detection theory: A user's guide* (2nd ed.). Mahwah, NJ: Lawrence Erlbaum Associates.

Marois, R., Yi, D., & Chun, M. M. (2004). The neural fate of consciously perceived and missed events in the attentional blink. *Neuron*, *41*(3), 465–472.

Mottaghy, F. M., Gangitano, M., Sparing, R., Krause, B. J., & Pascual-Leone, A. (2002). Segregation of areas related to visual working memory in the prefrontal cortex revealed by rTMS. *Cerebral Cortex*, *12*(4), 369–375.

Münchau, A., Bloem, B. R., Irlbacher, K., Trimble, M. R., & Rothwell, J. C. (2002). Functional connectivity of human premotor and motor cortex explored with repetitive transcranial magnetic stimulation. *Journal of Neuroscience*, *22*(2), 554–561.

Pascual-Leone, A., Valls-Solé, J., Wassermann, E. M., & Hallett, M. (1994). Responses to rapid-rate transcranial magnetic stimulation of the human motor cortex. *Brain*, *117*(4), 847–858.

Persaud, N., McLeod, P., & Cowey, A. (2007). Post-decision wagering objectively measures awareness. *Nature Neuroscience*, *10*(2), 257–261.

Pollen, D. A. (1995). Cortical areas in visual awareness. *Nature*, *377*(6547), 293–295.

Rees, G., Kreiman, G., & Koch, C. (2002). Neural correlates of consciousness in humans. *Nature Reviews Neuroscience*, *3*(4), 261–270.

Rounis, E., Stephan, K. E., Lee, L., Siebner, H. R., Pesenti, A., Friston, K. J., et al. (2006). Acute changes in frontoparietal activity after repetitive transcranial magnetic stimulation over the dorsolateral prefrontal cortex in a cued reaction time task. *Journal of Neuroscience*, *26*(38), 9629–9638.

Ruff, C. C., Blankenburg, F., Bjoertomt, O., Bestmann, S., Freeman, E., Haynes, J., et al. (2006). Concurrent TMS-fMRI and psychophysics reveal frontal influences on human retinotopic visual cortex. *Current Biology*, *16*(15), 1479–1488.

Sahraie, A., Weiskrantz, L., Barbur, J. L., Simmons, A., Williams, S. C., & Brammer, M. J. (1997). Pattern of neuronal activity associated with conscious and unconscious processing of visual signals. *Proceedings of the National Academy of Sciences of the United States of America*, *94*(17), 9406–9411.

Slachevsky, A., Pillon, B., Fourneret, P., Pradat-Diehl, P., Jeannerod, M., & Dubois, B. (2001). Preserved adjustment but impaired awareness in a sensory-motor conflict following prefrontal lesions. *Journal of Cognitive Neuroscience*, *13*(3), 332–340.

Slachevsky, A., Pillon, B., Fourneret, P., Renié, L., Levy, R., Jeannerod, M., et al. (2003). The prefrontal cortex and conscious monitoring of action: An experimental study. *Neuropsychologia*, *41*(6), 655–665.

Stefan, K., Gentner, R., Zeller, D., Dang, S., & Classen, J. (2008). Theta-burst stimulation: Remote physiological and local behavioral after-effects. *NeuroImage*, *40*(1), 265–274.

Swets, J. A. (1986). Indices of discrimination or diagnostic accuracy: Their ROCs and implied models. *Psychological Bulletin*, *99*(1), 100–117.

Szczepanowski, R., & Pessoa, L. (2007). Fear perception: Can objective and subjective awareness measures be dissociated? *Journal of Vision*, *7*(4), 10.

Talelli, P., Greenwood, R. J., & Rothwell, J. C. (2007). Exploring theta burst stimulation as an intervention to improve motor recovery in chronic stroke. *Clinical Neurophysiology*, *118*(2), 333–342.

Turatto, M., Sandrini, M., & Miniussi, C. (2004). The role of the right dorsolateral prefrontal cortex in visual change awareness. *NeuroReport*, *15*(16), 2549–2452.

Weiskrantz, L. (1997). *Consciousness lost and found: A neuropsychological exploration*. New York: Oxford University Press.

COGNITIVE NEUROSCIENCE, 2010, 1 (3), 176–183

Cortical reactivity and effective connectivity during REM sleep in humans

M. Massimini[1], F. Ferrarelli[2], M. J. Murphy[2], R. Huber[3], B. A. Riedner[2], S. Casarotto[1], and G. Tononi[2]

[1]University of Milan, Milan, Italy,
[2]University of Wisconsin, Madison, WI, USA,
[3]University of Zurich, Zurich, Switzerland

We recorded the electroencephalographic (EEG) responses evoked by transcranial magnetic stimulation (TMS) during the first rapid eye movement (REM) sleep episode of the night and compared them with the responses obtained during previous wakefulness and non-REM (NREM) sleep. Confirming previous findings, upon falling into NREM sleep, cortical activations became more local and stereotypical, indicating a significant impairment of the intracortical dialogue. During REM sleep, a state in which subjects regain consciousness but are almost paralyzed, TMS triggered more widespread and differentiated patterns of cortical activation that were similar to the ones observed in wakefulness. Similarly, TMS/high-density EEG may be used to probe the internal dialogue of the thalamocortical system in brain-injured patients that are unable to move and communicate.

Keywords: Consciousness; Sleep; Dream; Transcranial magnetic stimulation; EEG.

INTRODUCTION

According to a recent proposal (Tononi, 2004, 2008), consciousness depends critically not so much on firing rates, synchronization at specific frequency bands, or sensory input *per se*, but rather on the brain's ability to integrate information, which is contingent on the effective connectivity among functionally specialized regions of the thalamocortical system. Effective connectivity refers to the ability of a set of neuronal elements to causally affect the firing of other neuronal groups within a system (Lee, Harrison, & Mechelli, 2003). To test this proposal, our group has recently investigated whether the reduction of consciousness that occurs during non-rapid eye movement (NREM) sleep early in the night (Hobson & Pace-Schott, 2002; Stickgold, Malia, Fosse, Propper, & Hobson, 2001)

may be associated with changes in cortical effective connectivity (Massimini et al., 2005). To this end, transcranial magnetic stimulation (TMS) was combined with high-density electroencephalography (hd-EEG) to explore how the activation of one cortical area (the premotor area) is transmitted to the rest of the brain during wakefulness and NREM sleep. Results showed that, during quiet wakefulness, TMS triggered an initial response at the stimulation site that was followed by a sequence of fast waves (15–30 Hz) that moved to connected cortical areas several centimeters away. Upon entering NREM sleep Stages 3/4, the brain's response to TMS became a single large positive–negative slow wave that rapidly extinguished and did not propagate beyond the stimulation site. Based on this finding, we hypothesized that a significant breakdown of cortical effective connectivity may

Correspondence should be addressed to: G. Tononi, Department of Psychiatry, University of Wisconsin, 6001 Research Park Blvd, Madison, WI 53719, USA. E-mail: gtononi@wisc.edu

This work was supported by a National Institutes of Health Pioneer Award (to GT) and by the European Union (LSHM-CT-2005-518189 to MM).

www.psypress.com/cognitiveneuroscience
DOI: 10.1080/17588921003731578

actually explain why subjects awakened during NREM sleep early in the night report little or no conscious experience.

Of course, blank reports upon awakening from sleep are not the rule, and many awakenings yield dream reports (Casagrande, Violani, Lucidi, Buttinelli, & Bertini, 1996; Fagioli, 2002; Stickgold et al., 2001). Dreams can be at times as vivid and intensely conscious as waking experiences. Dream-like consciousness occurs during various phases of sleep, such as at sleep onset, during the last part of the night and, especially, during rapid eye movement (REM) sleep. Here we report the results of experiments in which we were able to record TMS-evoked potentials during the first REM sleep episode of the night and to compare them with the responses obtained during previous wakefulness and NREM sleep. During the transition from NREM to REM, while subjects were behaviorally asleep, the brain's reaction to TMS recovered fast oscillatory components and became similar to the one obtained during wakefulness, especially during the first 100–150 ms post-stimulus. The resurgence of fast waves was also associated with a partial recovery of cortical effective connectivity. These results suggest that directly measuring cortical reactivity and effective connectivity with TMS/hd-EEG may help in evaluating the brain's capacity for conscious experience in the absence of overt behavior, such as in sleeping subjects or in non-communicative, brain-injured patients.

METHODS

Subjects

Ten subjects (age 21–34, 3 females) were involved in the study. All participants gave written informed consent and the experiment was approved by the University of Wisconsin Human Subjects Committee. Prior to the experiment a neurological screening was performed to exclude potential adverse effects of TMS. The data presented here are from the five subjects in whom at least 30 trials of TMS-evoked potentials could be recorded during REM sleep.

TMS targeting

Cortical TMS targets were identified on T1-weighted magnetic resonance (MR) images (resolution 0.5 mm) of the subjects' whole heads acquired with a 3 T GE Signa scanner. In order to ensure precision and reproducibility of stimulation, we employed a navigated

brain stimulation (NBS) system (Nexstim Ltd, Helsinki, Finland). The NBS device located (with an error <3 mm) the relative positions of the subject's head and of the TMS coil by means of an optical tracking system. NBS also calculated the distribution and strength of the intracranial electric field induced by TMS. The coordinates of stimulation were input to a software aiming tool that ensured throughout the session the reproducibility of position, direction, and angle of the stimulator. We targeted the rostral part of the right premotor cortex (r-PM), 1 cm ahead of the anterior commissure and 2 cm lateral to the midline. This area is accessible through a central scalp position, far from any major head or facial muscle.

Stimulation parameters

Stimulation was performed by means of a Magstim figure-of-eight coil (model P/N9925), with a wing diameter of 70 mm, connected to a Magstim Rapid biphasic stimulator. Resting motor threshold (rMT) was measured prior to the experiment as the lowest stimulator output at which at least 5 out of 10 pulses, delivered on the optimal hand motor area, resulted in a motor evoked potential of 50 μV or greater in the abductor pollicis brevis (2). We stimulated r-PM at 110% of rMT, corresponding to a stimulator output between 54% and 67% and to a maximum estimated electric field on the target between 90 and 120 V/m. TMS was delivered with an interstimulus interval jittering randomly between 2 and 2.3 s.

EEG recordings

We recorded the spontaneous and TMS-evoked EEG by means of a 60 carbon electrodes cap and a specifically designed TMS-compatible amplifier (Nexstim). The artifact induced by TMS was gated and saturation of the amplifier was avoided by means of a proprietary sample-and-hold circuit that kept the analog output of the amplifier constant from 100 μs pre- to 2 ms post-stimulus (Virtanen, Ruohonen, Naatanen, & Ilmoniemi, 1999). To further optimize TMS compatibility, the impedance at all electrodes was kept below 3 kΩ. The EEG signals, referenced to an additional electrode on the forehead, were filtered (0.1–500 Hz) and sampled at 1450 Hz with 16 bit resolution. Four extra sensors were used to record the electrooculogram (EOG) and the chin electromyogram (EMG). In most cases, no signs of TMS-induced magnetic artifact were detected and in all cases the EEG signals were artifact-free from <10 ms post-stimulus. At the end of each experiment, a pen visible to the infrared camera was used to digitize the EEG electrode positions on the subject's head. During EEG recording, subjects' perception of the clicks produced by TMS coil's discharge was eliminated by means of inserted earplugs continuously playing a masking noise. Volume of the masking noise (always below 90 dB) was adjusted immediately prior to the experiment until the subjects reported that the TMS click was not perceptible. A thin layer of foam was placed between coil and scalp (resulting in less than 1 mm thickness when coil was pressed against the head) in order to attenuate bone conduction. As previously demonstrated, this procedure effectively prevented any contamination of EEG signals by auditory potentials elicited by TMS-associated clicks (Massimini et al., 2005, 2007).

General experimental procedures

Subjects were invited to the lab in the evening (9:30 p.m.) and were requested to stay awake until 1:00 a.m. After setting up the EEG cap and selecting the TMS target, recordings started at around 2:00 a.m. We chose this timing in order to extend the recordings to a circadian phase (early morning) that is more favorable to the occurrence of REM sleep (Dijk & Czeisler, 1995). During the experiment subjects were lying eyes-closed on a reclining chair with a head-rest that allowed a comfortable and stable head position. Participants were requested to stay awake while at least 200 trials were recorded. Afterwards, we allowed the subjects to fall asleep while the spontaneous EEG was monitored and TMS was continuously delivered. In order to avoid overheating of the coil, we interrupted and restarted the stimulation according to the EEG pattern of interest. Responses during NREM sleep were collected in 7 out of 10 subjects. Five of these subjects subsequently entered REM sleep for at least 2 min and were included in the analysis. Throughout the recording session the stability of stimulation coordinates was continuously monitored. If a movement of >5 mm occurred, the session was interrupted and the coil was repositioned. At the end of the experiment, the stimulation coordinates were recorded and the electrode positions were digitized.

Dream reports

In order to evaluate subjective experience we collected a dream report in all the subjects ($n = 5$) who entered REM sleep. Two subjects awakened spontaneously during REM sleep while the remaining three were awakened as soon as the REM sleep episode was judged to be over, based on the online inspection of the EEG, EOG, and EMG traces. Upon awakening subjects were asked to remember what was going in their mind in the time prior to waking (Stickgold et al., 2001). All reports were collected by dictation into a portable digital audio recorder (Olympus DS-50). Recordings were transcribed and subsequently edited for total recall count (TRC) (Antrobus, 1983), removing extraverbal utterances, nonwords, repeated words, and secondary elaborations. TRCs were obtained using the word-count function of Microsoft Word.

Data analysis

Data analysis was performed in Matlab (The Matworks, MA) and Curry 5.0 (Philips GmbH, Germany). Sleep stages were scored offline according to Rechtschaffen and Kales (1968) on two EEG channels (C3, C4) re-referenced to the mastoid, one EOG channel and one EMG channel. We classified the single trials according to the stages during which they were collected. Trials containing noise, muscle activity, or eye movements were rejected. The trials collected during wakefulness, NREM sleep Stages 3–4, and REM sleep were averaged separately. The averaged signals were baseline corrected and bandpass filtered between 2 and 100 Hz. In all subjects, the averaged TMS-evoked potentials was characterized, during wakefulness, by two consecutive positive peaks (peak 1 and peak 2) occurring at around 20 and 60 ms, respectively (Figure 2). We characterized changes in the shape of the early EEG response to TMS by detecting significant changes of the amplitude of peak 1 and peak 2 across states of vigilance. To this aim, we measured, on each single-trial response, the maximum voltage at around (±10 ms) the latency of peak 1 and peak 2 (as detected during wakefulness). After that, we performed a Student's t-test on the single-trial amplitude distributions obtained in the different conditions.

In order to perform source modeling, the averaged responses, the MR imaging (MRI) sets and the electrode coordinates were input to the software package Curry 5.0. Following a semi-automatic segmentation of the individual MRI images, we implemented a boundary element model (BEM) of the head having three compartments of fixed conductivities (scalp: 0.33 S/m; skull: 0.0042 S/m; brain: 0.33 S/m). The cortical surface was also reconstructed with a 6 mm resolution and modeled with approximately 14,000 rotating dipoles. Next, the electrode positions were projected onto the skin surface and the lead field matrix was calculated. We estimated the current density on the cortical surface using the minimum norm least squares (L2 Norm) method (weighted for depth bias removal) (Hamalainen & Ilmoniemi, 1994). To identify the cortical areas involved by TMS-evoked activity we proceeded as follows: (i) we calculated the global mean field power (GMFP; Hamalainen & Ilmoniemi, 1994) as the mean of the absolute voltage recorded from electrodes in average reference; (ii) to identify the latencies where TMS evoked a significant response, we detected the time intervals where the post-stimulus amplitude of the GMFP was 3 standard deviations larger than pre-stimulus activity; (iii) at these time points, we estimated the location of maximum neural activity using the L2 norm. Thus, we estimated the location of the maximum activation (top 10%) at each significant time sample, using Matlab 7.0. The identified sources were then plotted on the individual cortical surface and color-coded according to the latency of their involvement.

RESULTS

While TMS was delivered, the following percentages of sleep stages were obtained across the five subjects who entered REM sleep: wakefulness, 45.4%; Stage 1, 9.1%; Stage 2, 15.6%; Stages 3–4, 22.6%; REM, 7.3%. In all cases, at least 150 single-trial evoked responses could be retained during wakefulness and NREM sleep. In four subjects, all the trials averaged in the NREM sleep condition were collected during Stages 3 and 4, while in one subject trials collected during NREM Stages 2, 3 and 4 had to be averaged together in order to obtain a stable NREM response. TMS/hd-EEG measurements during REM sleep were more challenging. Indeed, this sleep stage tends to occur unpredictably and is rather unstable, especially during the first sleep episode. As a consequence, periods of clear-cut REM sleep were much shorter. Specifically, we obtained, across all subjects, the following numbers of seconds (and of TMS-evoked single trials) during REM sleep: 132 s (42 trials), 163 s (53 trials), 169 s (52 trials), 230 s (92 trials), 348 s (145 trials). While TMS-evoked potentials have an excellent signal-to-noise ratio, a limited number of trials hampers the reliability of the average response, especially at late latencies, where intertrial variability increases. Thus, to assess the reliability of the overall scalp-recorded average response at different latencies, we detected the time-samples where the GMFP was significant (post-stimulus amplitude > of 3 standard deviations of pre-stimulus; see also "Methods" and

Massimini et al., 2005). In each subject, this test was repeated for each condition (including wakefulness, NREM sleep) using the maximum number of trials collected during REM sleep. In all subjects, a significant response could be detected during the first 100-150 ms, while in the subject in whom 145 single trials could be analyzed the significant response extended to 300 ms. For this reason, the analysis of TMS-evoked potentials was restricted to the first 150 ms in all cases except for this subject, in whom source modeling was also performed. All subjects, prompted on awakening from REM sleep, provided a dream report. TWC varied largely across subjects (ranging from 45 words to 237 words) and tended to correlate ($p < .1$) with the length of the period that subjects spent in REM sleep.

In all vigilance states, TMS elicited a time-locked response that could be appreciated on a single-trial basis. Figure 1A displays the single-trial responses recorded from one electrode located under the stimulator (corresponding to FC2, in the 10–20 system) during a transition from wakefulness through Stage 1 and NREM to REM sleep. In this representation, signals are bandpassed from 15 to 100 Hz, in order to highlight the presence of fast oscillations in the single-trial responses. Figure 1B shows the corresponding averages (filtered 2–100 Hz) calculated in the four vigilance states. During wakefulness, TMS triggered a sustained response made of a sequence of time-locked, high-frequency (20–35 Hz) oscillations in the first 100 ms. As soon as the subject transitioned into Stage 1, the amplitude of the first positive component (between 0 and 40 ms) increased by 40%, while the subsequent waves were dampened. During NREM sleep the response changed markedly; the first positive component became larger (120% increase), slower, and was followed by a negative rebound after which the response extinguished. Upon transitioning to REM sleep the cortical response to TMS recovered wakefulness-like fast oscillations, despite the subject being behaviorally asleep.

Figure 2 shows that the resumption of fast oscillations during REM sleep was reproducible across subjects and, at the same time, it highlights significant differences between the REM sleep and the wakefulness response. In REM sleep, compared to wakefulness, the first positive component (peak 1) was larger ($p < .05$) and the second (peak 2) was smaller ($p < .05$) in each subject (Student's t-test, comparing single-trial amplitude distributions). During NREM sleep, peak 1 was larger compared to both REM sleep ($p < .001$) and wakefulness ($p < .001$), while peak 2 was replaced by a negative wave.

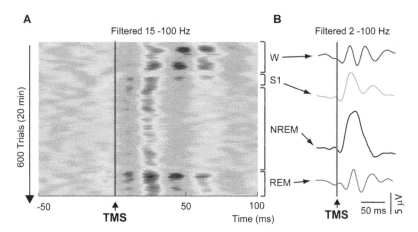

Figure 1. (A) Single trial TMS-evoked responses are recorded from one channel located under the stimulator (FC2) while a subject transitions from wakefulness (W), through sleep Stage 1 (S1) and NREM sleep (NREM) to REM sleep. The red line marks the time of TMS. Single-trial EEG data are bandpass filtered (15 to 100 Hz) and color coded for voltage (red for positive, blue for negative). (B) The averaged responses (filtered from 2 to 100 Hz) calculated in the four vigilance states are depicted. The onset of REM sleep is associated with a resumption of TMS-evoked fast oscillations.

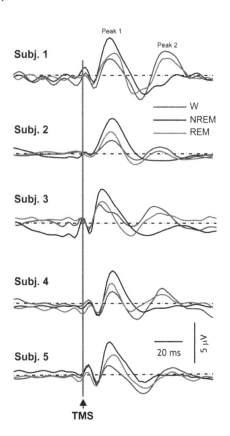

Figure 2. The averaged responses obtained in all subjects during wakefulness (blue traces), NREM sleep (black traces) and REM sleep (red traces) are compared. The vertical red line marks the time of TMS. TMS-evoked potentials undergo systematic changes across states of vigilances.

In one subject, we were able to collect a sufficient number of trials to perform source modeling and to compare REM sleep to NREM sleep and wakefulness

also at longer latencies (Figure 3). This analysis revealed that the complex pattern of long-lasting, long-range activation triggered by TMS during wakefulness broke down during NREM sleep and that it partially recovered, within the first 150 ms post-stimulus, during REM sleep. Figure 3 also shows that long-latency components that were present in wakefulness were obliterated during REM sleep. Unfortunately, due to the short duration of clear-cut REM sleep periods, the long latency response could not be analyzed in the other subjects.

DISCUSSION

The level and quality of conscious experience can vary dramatically across the sleep–wake cycle. During NREM sleep early in the night, consciousness can nearly vanish (Hobson & Pace-Schott, 2002; Pivik & Foulkes, 1968; Suzuki et al., 2004) despite persistent neural activity in the thalamocortical system (Steriade, Timofeev, & Grenier, 2001). Why is this so? The present work confirms the results of previous measurements (Massimini et al., 2005) by showing that, during NREM sleep Stages 3 and 4 early in the night, thalamocortical circuits remain active and reactive but lose their ability to interact and to produce complex, integrated responses. Indeed, in this state, TMS failed to trigger a sustained, long-range pattern of activation and instead evoked a simple positive–negative wave that remained local. Interestingly, this stereotypical response has been shown to share fundamental characteristics with the slow waves that occur spontaneously during NREM

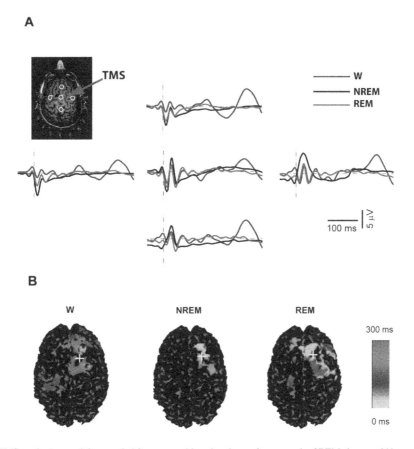

Figure 3. (A) The TMS-evoked potentials recorded from one subject, in whom a long stretch of REM sleep could be recorded, are displayed (blue: wakefulness, black: NREM sleep, red: REM sleep). The traces were recorded from the channels indicated by red dots in the upper left panel, where the site of stimulation on the subject's MRI is indicated by the red arrow. (B) Spatiotemporal cortical maps of TMS-evoked cortical activation during wakefulness, NREM, and REM sleep. For each significant time sample, maximum current sources were plotted and color-coded according to their latency of activation (light blue, 0 ms; red, 300 ms). The yellow cross marks the TMS target on the cortical surface. During REM sleep, the resumption of TMS-evoked fast oscillations was associated with a partial recovery of cortical effective connectivity.

sleep (Massimini et al., 2007). This evidence suggests that the mechanisms underlying the generation of slow waves may also be responsible for blocking the emergence of specific long-range responses during NREM sleep (Massimini, Tononi, & Huber, 2009; Tononi & Massimini, 2008). Upon falling asleep, due to a dampening of brainstem noradrenergic, serotoninergic, and cholinergic activating systems (Steriade, 2004), cortical neurons become bistable and inevitably tend to fall into a silent, hyperpolarized state (down-state) after a period of activation (up-state) (Timofeev, Grenier, & Steriade, 2001). This mechanism provides the mechanism for the slow oscillations of sleep, where large populations of cortical neurons spontaneously alternate between up- and down-states (Hill & Tononi, 2005). In addition, bistability may also be contributed for by a shift in the balance of synaptic excitation and inhibition toward inhibition due to changes in the neuromodulatory

milieu (Esser, Hill, & Tononi, 2009). In any case, bistability may prevent the emergence of sustained, complex thalamocortical interactions. Thus, in this condition, any local activation, whether occurring spontaneously or induced by a stimulus (such as TMS), will converge into a silent neuronal down-state and into a stereotypical EEG slow wave. While this seems to occur at least during deep NREM sleep early in the night (Tononi & Massimini, 2008), it is hard to predict what would happen during REM sleep. In this state, while noradrenergic and serotoninergic arousing systems remain silent, brainstem cholinergic neurons come back to activity (Pace-Schott & Hobson, 2002), spontaneous slow waves disappear, and the EEG becomes, at least superficially, similar to the one of wakefulness. However, despite this apparent resemblance with wakefulness, it is difficult to infer the degree of underlying thalamocortical bistability during

REM sleep based on the presence of a low-voltage EEG alone. Indeed, it was shown that, during NREM sleep, TMS delivered during short stretches of low-amplitude EEG sleep was still able to trigger full-fledged slow waves (Massimini et al., 2007). The present experiments demonstrate that, during REM sleep, TMS evokes a response that is more similar to the one observed in wakefulness, namely, a sequence of fast oscillations (Figures 1 and 2) occurring in the first 150 ms. This evidence suggests that the discharge of mesopontine cholinergic neurons alone may represent a major excitatory input that can largely prevent the emergence of thalamocortical bistability during REM sleep.

Despite obvious similarities, the REM and the wakefulness responses also showed consistent differences; in REM sleep, the first positive component (peak 1) was always larger and the second one (peak 2) was always smaller compared to wakefulness (Figure 2). In this sense, the REM response tended to share some features with the NREM sleep response, where peak 1 reached its maximum amplitude and peak 2 disappeared. Notably, the REM sleep response recorded in the present experiment strongly resembled the TMS-evoked potential obtained during sleep Stage 1 in a previous work (Massimini et al., 2005), suggesting that Stage 1 and REM sleep may be supported by a similar degree of cortical activation.

Source modeling revealed that, as in wakefulness, the resumption of fast oscillations during REM sleep was associated with a pattern of activation that was more complex and widespread than one of NREM sleep. This observation, although limited to a single subject, corroborates the hypothesis that cortical effective connectivity may play a role in the shifts of conscious experience that occur during sleep. Indeed, the subject in Figure 3 spent the longest time in REM sleep (348 s) and reported a long dream recall (237 words) upon awakening. The persistence, to some degree, of long-range cortico-cortical effective connectivity has been also reported during Stage 1 (see Figure S2 in Massimini et al., 2005), another sleep stage associated with frequent and long dream reports (Foulkes, 1966). Figure 3 also shows that long-latency components that were present in wakefulness were obliterated during REM sleep. This finding is consistent with the notion that late components of peripherally evoked potentials are also dampened during REM sleep (Goff, Allison, Shapiro, & Rosner, 1966; Wesensten & Badia, 1988). In future works it would be interesting to systematically collect TMS/hd-EEG measures of thalamocortical bistability and effective connectivity during the whole night and to correlate them with dream reports. This approach might represent a valid attempt to study the neural correlates of consciousness during sleep on a finer time scale, beyond the REM/NREM sleep dichotomy and beyond traditional sleep staging. Similarly, exploring brain activity on a finer spatial scale with functional neuroimaging has already shed light on the neural mechanisms underlying changes in the quality of consciousness across the sleep–wake cycle (Maquet et al., 1996, 2005).

Besides their possible relevance for sleep physiology, the present experiments demonstrate that TMS/hd-EEG may represent an effective way to probe the internal dialogue of the thalamocortical system in the absence of any behavioral cue. TMS triggered more widespread and differentiated patterns of EEG activation, just as it does during normal wakefulness, upon entering REM sleep, a state in which subjects are conscious but almost paralyzed. Hence, in the future, this technique may be employed as an aid to evaluate the brain's capacity for consciousness in brain-injured patients who are unable to move and communicate (Massimini, Boly, Casali, Rosanova, & Tononi, 2009).

REFERENCES

Antrobus, J. (1983). REM and NREM sleep reports: Comparison of word frequencies by cognitive classes. *Psychophysiology*, *20*(5), 562–568.

Casagrande, M., Violani, C., Lucidi, F., Buttinelli, E., & Bertini, M. (1996). Variations in sleep mentation as a function of time of night. *International Journal of Neuroscience*, *85*(1–2), 19–30.

Dijk, D. J., & Czeisler, C. A. (1995). Contribution of the circadian pacemaker and the sleep homeostat to sleep propensity, sleep structure, electroencephalographic slow waves, and sleep spindle activity in humans. *Journal of Neuroscience*, *15*(5, Pt 1), 3526–3538.

Esser, S. K., Hill, S., & Tononi, G. (2009). Breakdown of effective connectivity during slow wave sleep: Investigating the mechanism underlying a cortical gate using large-scale modeling. *Journal of Neurophysiology*, *102*(4), 2096–2111.

Fagioli, I. (2002). Mental activity during sleep. *Sleep Medical Review*, *6*(4), 307–320.

Foulkes, W. D. (1962). Dream reports from different stages of sleep. *Journal of Abnormal and Social Psychology*, *65*, 14–25.

Goff, W. R., Allison, T., Shapiro, A., & Rosner, B. S. (1966). Cerebral somatosensory responses evoked during sleep in man. *Electroencephalography and Clinical Neurophysiology*, *21*(1), 1–9.

Hamalainen, M. S., & Ilmoniemi, R. J. (1994). Interpreting magnetic fields of the brain: Minimum norm estimates. *Medical and Biological Engineering and Computing*, *32*(1), 35–42.

Hill, S., & Tononi, G. (2005). Modeling sleep and wakefulness in the thalamocortical system. *Journal of Neurophysiology, 93*(3), 1671–1698.

Hobson, J. A., & Pace-Schott, E. F. (2002). The cognitive neuroscience of sleep: Neuronal systems, consciousness and learning. *Nature Reviews Neuroscience, 3*(9), 679–693.

Lee, L., Harrison, L. M., & Mechelli, A. (2003). A report of the functional connectivity workshop, Dusseldorf 2002. *NeuroImage, 19*(2, Pt 1), 457–465.

Maquet, P., Peters, J., Aerts, J., Delfiore, G., Degueldre, C., Luxen, A., et al. (1996). Functional neuroanatomy of human rapid-eye-movement sleep and dreaming. *Nature, 383*(6596), 163–166.

Maquet, P., Ruby, P., Maudoux, A., Albouy, G., Sterpenich, V., Dang-Vu, T., et al. (2005). Human cognition during REM sleep and the activity profile within frontal and parietal cortices: A reappraisal of functional neuroimaging data. *Progress in Brain Research, 150*, 219–227.

Massimini, M., Boly, M., Casali, A., Rosanova, M., & Tononi, G. (2009). A perturbational approach for evaluating the brain's capacity for consciousness. *Progress in Brain Research, 177*, 201–214.

Massimini, M., Ferrarelli, F., Esser, S. K., Riedner, B. A., Huber, R., Murphy, M., et al. (2007). Triggering sleep slow waves by transcranial magnetic stimulation. *Proceedings of the National Academy of Sciences of the United States of America, 104*(20), 8496–8501.

Massimini, M., Ferrarelli, F., Huber, R., Esser, S. K., Singh, H., & Tononi, G. (2005). Breakdown of cortical effective connectivity during sleep. *Science, 309*(5744), 2228–2232.

Massimini, M., Tononi, G., & Huber, R. (2009). Slow waves, synaptic plasticity and information processing: Insights from transcranial magnetic stimulation and high-density EEG experiments. *European Journal of Neuroscience, 29*(9), 1761–1770.

Pace-Schott, E. F., & Hobson, J. A. (2002). The neurobiology of sleep: Genetics, cellular physiology and subcortical networks. *Nature Reviews Neuroscience, 3*(8), 591–605.

Pivik, T., & Foulkes, D. (1968). NREM mentation: Relation to personality, orientation time, and time of night. *Journal of Consulting and Clinical Psychology, 32*(2), 144–151.

Rechtschaffen A., & Kales A. (1968). *A manual of standardized terminology techniques and scoring system for sleep stages of human subjects.* Public Health Service U. S. Government Printing Office, Washington, D.C.

Steriade, M. (2004). Acetylcholine systems and rhythmic activities during the waking–sleep cycle. *Progress in Brain Research, 145*, 179–196.

Steriade, M., Timofeev, I., & Grenier, F. (2001). Natural waking and sleep states: A view from inside neocortical neurons. *Journal of Neurophysiology, 85*(5), 1969–1985.

Stickgold, R., Malia, A., Fosse, R., Propper, R., & Hobson, J. A. (2001). Brain–mind states: I. Longitudinal field study of sleep/wake factors influencing mentation report length. *Sleep, 24*(2), 171–179.

Suzuki, H., Uchiyama, M., Tagaya, H., Ozaki, A., Kuriyama, K., Aritake, S., et al. (2004). Dreaming during non-rapid eye movement sleep in the absence of prior rapid eye movement sleep. *Sleep, 27*(8), 1486–1490.

Timofeev, I., Grenier, F., & Steriade, M. (2001). Disfacilitation and active inhibition in the neocortex during the natural sleep–wake cycle: An intracellular study. *Proceedings of the National Academy of Sciences of the United States of America, 98*(4), 1924–1929.

Tononi, G. (2004). An information integration theory of consciousness. *BMC Neuroscience, 5*, 42.

Tononi, G. (2008). Consciousness as integrated information: A provisional manifesto. *Biological Bulletin, 215*(3), 216–242.

Tononi, G., & Massimini, M. (2008). Why does consciousness fade in early sleep? *Annals of the New York Academy of Sciences, 1129*, 330–334.

Virtanen, J., Ruohonen, J., Naatanen, R., & Ilmoniemi, R. J. (1999). Instrumentation for the measurement of electric brain responses to transcranial magnetic stimulation. *Medical and Biological Engineering and Computing, 37*(3), 322–326.

Wesensten, N. J., & Badia, P. (1988). The P300 component in sleep. *Physiology and Behavior, 44*(2), 215–220.

COGNITIVE NEUROSCIENCE, 2010, 1 (3), 184–192

Feeling in control of your footsteps: Conscious gait monitoring and the auditory consequences of footsteps

Fritz Menzer[1], Anna Brooks[1,2], Pär Halje[1], Christof Faller[1], Martin Vetterli[1], and Olaf Blanke[1]

[1]Ecole Polytechnique Fédérale de Lausanne, Lausanne, Switzerland
[2]Southern Cross University, Coffs Harbour, Australia

A fundamental aspect of the "I" of conscious experience is that the self is experienced as a single coherent representation of the entire, spatially situated body. The purpose of the present study was to investigate agency for the entire body. We provided participants with performance-related auditory cues and induced online sensorimotor conflicts in free walking conditions investigating the limits of human consciousness in moving agents. We show that the control of full-body locomotion and the building of a conscious experience of it are at least partially distinct brain processes. The comparable effects on agency using audio-motor and visuo-motor cues as found in the present and previous agency work may reflect common supramodal mechanisms in conscious action monitoring. Our data may help to refine the scientific criteria of selfhood and are of relevance for the investigation of neurological and psychiatric patients with disturbance of selfhood.

Keywords: Motor awareness; Self; Agency; Sensorimotor; Auditory; Gait; Consciousness.

INTRODUCTION

In a scene from *The Man Who Knew Too Much* (Alfred Hitchcock, 1956), Dr Ben McKenna played by Jimmy Stewart walks down an empty street in London. He is searching for his son Hank who he believes was kidnapped by a couple in Morocco two days earlier. Dr McKenna is alone and suspects that people may be following him. Actor and audience hear footsteps that do not seem to come from Dr McKenna's feet. However, the heard footsteps slow down, stop, accelerate as Dr McKenna slows down, stops, and accelerates. Where are the footsteps coming from? Is someone following Dr McKenna? Do they reflect a loud and distorted echo of his footsteps? Dr McKenna

turns around several times during that sequence but sees nobody else. Determining whether we are causing the perceptual events that we perceive or whether somebody or something else causes these events is an important function and is generally referred to as agency. How do I know whether I am causing the rhythmic sounds of footsteps? When do I begin to suspect that another agent is causing the sounds I hear?

"Actions are critical steps in the interaction between the self and the external milleu" (Jeannerod, 2007) and may reveal—especially when self-generated and not responses to external events—the intentions, states, and goals of the acting self. Moreover, such self-generated actions also alter the perceptual environment.

Correspondence should be addressed to: Olaf Blanke, Laboratory of Cognitive Neuroscience, Brain-Mind Institute, Station 19, Ecole Polytechnique Fédérale de Lausanne (EPFL), 1015 Lausanne, Switzerland. E-mail: olaf.blanke@epfl.ch

www.psypress.com/cognitiveneuroscience DOI: 10.1080/17588921003743581

Previous work has shown that agency judgments can be influenced by the manipulation of perceptual and sensorimotor cues during different phases of action execution (Farrer et al., 2008; Fourneret & Jeannerod, 1998; Frith, Blakemore, & Wolpert, 2000; Knoblich & Kircher, 2004; Sato & Yasuda, 2005). Thus, it has been tested whether discrepancies between sensory predictions and the actual sensory input (re-afference) may lead participants to judge self-generated actions as externally generated. Following early work by Nielsen (1963), this latter form of agency (or conscious action monitoring) has recently been the topic of intensive research (Fourneret & Jeannerod, 1998; Franck et al., 2001). In these studies participants' agency was measured in response to varied sensorimotor incongruencies between visual and motor (and proprioceptive) signals (that are generally congruent during action execution). In these studies the authors investigated sensorimotor conflicts by manipulating the visual position of the participant's hand as seen on a computer screen and the participant's actual hand while a simple goal-directed motor task was carried out. Participants were asked to direct their hand to a specified target position. Direct vision of the participant's hand during the target-directed action was occluded and visual feedback of the arm movements (again as seen on a computer screen) was systematically deviated from its actual movement path. These studies showed that participants automatically aligned their hand trajectories with a visual target on the computer screen while compensating for a displayed spatial deviation. The participants were often unaware of their online movement corrections and judged many of these movements as non-deviated; these data also revealed that agency decreased with increasing sensorimotor incongruency between these visual and motor cues. In a similar experimental set-up, Franck et al. (2001) investigated the influence of temporal cues on agency judgments. The authors introduced different temporal delays between the visual position of the participant's hand (as seen on a computer screen) and the participant's actual hand while the same motor task was carried out. As observed for spatial deviations, increasing delay durations between visual and sensorimotor cues were found to decrease agency (see also Shimada, Qi, & Hiraki, 2010).

Most previous work has focused on the investigation of performance-related visual cues on agency (Daprati et al., 1997; Farrer et al., 2003, 2008; Fourneret & Jeannerod, 1998; Franck et al., 2001; Knoblich & Kircher, 2004; Shimada et al., 2010; Tsakiris, Haggard, Franck, Mainy, & Sirigu, 2005; van den Bos & Jeannerod, 2002). Auditory action consequences may differ in their effects from visual action consequences, as the auditory detection of temporal discrepancies may be an especially powerful agency cue due to the excellent timing of auditory perception. In spite of this such auditory effects on agency have only rarely been tested (Asai & Tanno, 2008; Knoblich & Repp, 2009; Repp & Knoblich, 2007; Sato & Yasuda, 2005).

The above example from the *The Man Who Knew Too Much* points to a fundamental difference between the conscious monitoring of actions as tested in most of the previous agency work and conscious monitoring related to questions such as "How do I know whether the footsteps I hear are mine?" Almost all previous work on agency has focused on the investigation of performance-related sensory cues for upper limb actions (actions of fingers, hands, or arms). Yet a fundamental aspect of the "I" of conscious experience is that the self is experienced as a single coherent representation of the entire, spatially situated body, not as several separate body parts (Blanke & Metzinger, 2009). Recently, several experimental procedures have been reported that allow one to test full-body ownership or the conscious experience of identifying with one's body and of being localized within one's body (Ehrsson, 2007; Lenggenhager, Mouthon, & Blanke, 2009; Lenggenhager, Tadi, Metzinger, & Blanke, 2007; Petkova & Ehrsson, 2008) and such changes in full-body ownership have also been shown to modify the perception of tactile stimuli (Aspell, Lenggenhager, & Blanke, 2009). Concerning agency and because the participants' body position was kept constant (except for actions of finger, hand, or arm), previous agency studies did not investigate this fundamental aspect of the global bodily self, because this requires movement of the entire body of the participant as during locomotion. Footstep-related signals during locomotion are probably one of the most common performance-related auditory cues (alongside speech or eating) and of significant relevance for the self. This is suggested by the scene from *The Man Who Knew Too Much* and by clinical data in neurological patients (i.e. Blanke, Ortigue, Coeytaux, Martory, & Landis, 2003). Moreover, walking differs from upper limb actions in several physiological ways: Gait is cyclic, more rarely immediately goal-directed, and is generally considered a highly automatic and unconscious action with important control centres in spinal cord and brainstem (Armstrong, 1988; Grillner & Wallen, 1985). Collectively, these data suggest that agency for the full body may differ from agency for the upper limb.

To investigate conscious action monitoring for the entire body we asked participants to make agency judgments during locomotion. For this we developed a portable device that allows the introduction of different systematic temporal delays between the participants' footsteps and the auditory consequences of those footsteps. This was combined with an analysis of the walking speed during the different tested delays. We predicted that an increase in delay should lead to a decrease in agency judgments as observed in previous arm agency studies, but that longer delays (those approaching the next footstep) would lead to increases in agency judgments.

METHODS

Participants

Eleven healthy participants with normal or corrected-to-normal hearing (five female; one left-handed), aged 21–30 years, volunteered for the experiment. Experiments were conducted in accordance with the Declaration of Helsinki and accepted by the local ethics commission (University Hospital of Lausanne).

Procedure

After a participant's shoes were removed and replaced with experimental shoes (see below), we taped the microphone chords to the participant's trousers and the microphones to their shoes, and the backpack in which the laptop computer running the experiment software was carried and the eyewear were donned (Figure 1). In order to minimize visual contributions to the task, participants were instructed to keep their heads up and not to inspect the patterns of movement of their limbs (controlled by experimenter). Participants also wore occluding eyewear such that visual information from below the head was impaired (field of view was limited to ~15° × ~15°).

After this, participants performed as many baseline trials as necessary to familiarize themselves with the experimental setup, walking, and task. They were instructed to walk continuously in a clockwise circuit prescribed by a cordoned-off hallway measuring 20 × 4 m (Figure 1). Individual trials were initiated by themselves via a button on a handheld wireless device. Two other buttons were assigned as "response" buttons. Two alternative forced-choice judgments were registered before further trials could commence. Partici-

 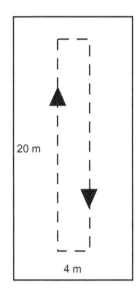

Figure 1. Experimental setup. (Left) A participant wearing the recording and stimulation system. (Right) Plan of the hallway (measuring 20 m × 4 m) where particpants were instructed to walk continuously in a clockwise circuit.

pants were instructed *not* to initiate trials on their approach to the turn at the end of the hallway, but rather to wait to initiate the next trial until the commencement of their walk down the length of the space (to minimize potential left–right differences in the auditory cues that would be associated with turning in a consistent (clockwise) direction). Participants were instructed to walk at a normal and relaxed speed at all times (as if they were "taking a stroll along a footpath") and informed that trials would be presented in four blocks, and that between those blocks they would have the opportunity to be seated and take a drink of water (break duration was determined by participant comfort; average duration of each block was 6 min 22 s, ±48 s; SD).

Blocks consisted of randomly selected trials from each auditory delay condition. The duration of each individual trial was 7 s. This allowed each participant to take on average 11.3 footsteps (±1.1 footsteps; SD) in each delay condition. After each trial participants were instructed to respond by "yes" or "no" as to whether the walking that they heard over the headphones corresponded to the walking they had just performed. We analyzed the percentage of "yes" responses (agency judgment). We studied audio-motor agency based on experimental paradigms and agency questions that are comparable to those employed in previous work on visuo-motor agency (Fourneret & Jeannerod, 1998; Franck et al., 2001; Kannape, Schwabe, Tadi, & Blanke, in press; Shimada et al., 2010).

Materials

The individual footsteps were recorded by microphones (Voice Technologies VT 500, Switzerland) that were attached to each shoe (tip; Figure 1). These recordings were presented binaurally, with a delay determined by the condition from which the trial was selected. Conditions were presented in randomized fashion and participants wore closed headphones with high ambient noise attenuation (BeyerDynamic DT 770M, Germany).

Auditory delays were implemented via custommade software on a portable computer (MacBook 2.16 GHz Core 2 Duo, USA), carried by participants in a backpack. We tested 19 different temporal delays ranging from 16 to 1800 ms that were implemented "online" while participants were walking. Each condition was repeated 20 times and represented a different delay between the actual footstep and the presented auditory cue associated with the actual footstep. The tested delays are shown in Figure 2. The minimal delay the system was able to provide was 16 ms. For each trial the gait period was determined from the

audio signal recorded by the microphones. The gait period was calculated as the position of the maximum of the autocorrelation of the signal's envelope outside of the range of 0–0.8 s (assuming that the gait period is >0.8 s). The custom software allowed for precision recording of the auditory profile of participants' footsteps over time. To exclude personal "shoe-specific" auditory signals, all participants wore the same pair of experimental shoes, adjusted for size via an ankle strap. A hand-held device (Wii Remote, Nintendo, Japan) was used to register, after each trial, the participant's agency judgment. For statistical analysis a within-subjects design was used.

RESULTS

Agency judgments

Data from the agency judgment task were collated across participants for each delay condition and are represented in Figure 2A, showing that agency judgments depend on delay. As predicted, the data reveal the highest percentage of confirmatory agency judgments (~90%) for delays of 16 and 100 ms which rapidly decrease to 34% and 28% for delays between 250 and 450 ms. For 450–750ms delays, the percentage of agency judgments increased continuously to ~75%. Over the next 750 ms this sinusoidal pattern was repeated (maximum of ~65% at 1300 ms). A 4 parameter damped sine wave model fitted to the data yields an R^2 value of .92, thus explaining 92% of the variance (significant fit; $p < .05$). These data provide evidence of a predictable relationship between agency judgments and the extent of the auditory delay.

We next analyzed whether the damping with increasing delays reflects increased variability of the peaks of maximal and minimal agency judgments for larger delays. Inspection of individual data (Figure 2B) shows that agency judgments were as precise (and as elevated) as for the minimal delay of 16 ms in most participants, showing that variability does not account for dampening. Moreover, the sinusoidal pattern suggests that the cyclic nature of walking (Murray, 1967; Blanc Balmer, C., Landis, T., & Vingerhoets,, 1999) interfered with agency judgments. Gait agency may depend not only on delay, but also on gait cycle and gait speed. All participants were instructed to walk at their habitual speed, but as stride length and walking speed depend on individualistic parameters (i.e., height and leg length; Macellari, Giacomozzi, & Saggini, 1999), we next analyzed our participants' walking speed.

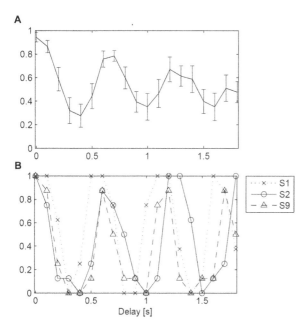

Figure 2. Uncorrected agency judgments. A. The plot shows agency judgments (mean ± 1 SD) across delay conditions. The data show that agency judgments depend on the delay (i.e. initial decrease in agency judgments until ~450 ms). The sinusoidal pattern suggests that the cyclic nature of walking also influences agency judgments. Note the dampening of the sinusoidal pattern with increasing delays. B. Individual data from three participants are shown (S1, S2, S9). These data show that agency judgments were as precise (and as elevated) as for the minimal delay of 16 ms. Note that there is no apparent dampening of the agency curve.

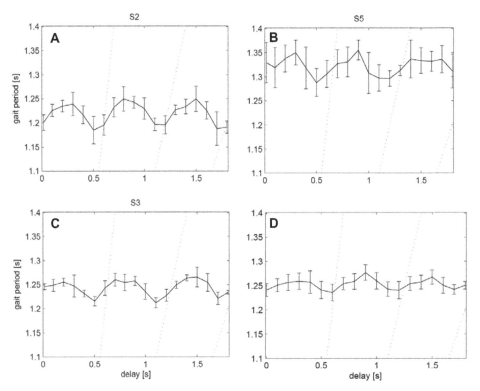

Figure 3. Gait period is shown as a function of delay condition for three individual subjects (A–C) and all participants (D). The average gait period showed small but systematic variation with the delay period in a sinusoidal pattern for the three depicted participants (S2, S3, S5). Because the participants with longer gait periods (higher curve on *y*-axis) also had slower varying gait period vs. delay curves (curve stretched along the *x*-axis), the average data across all participants (Figure 3D) were obtained by stretching the individual curves by the same factor in *x* and *y* direction until their mean was the same as the previously calculated mean gait period, and subsequently taking the mean over all stretched curves. Thus, before calculating the average curve, the curve of each participant was stretched along the dotted lines until their mean matched the average gait period across all participants. Note that the delay conditions with smallest gait period (or fastest walking speed) overlap with the delay conditions for which we found that agency judgments reached maximal values (compare with Figure 2A).

Gait-period and walking speed depend on delay

Analysing the gait period of each participant as a function of delay condition, we found that the average gait period was 1.25 s (0.13 s (*SD*); range: 1.01–1.79 s) and that this varied systematically with the delay period in a sinusoidal pattern (Figure 3). On average participants' footsteps were separated by 613 ms, compatible with physiological data in healthy subjects (Blanc, Balmer, Landis, & Vingerhoets, 1999; Macellari et al., 1999). Figure 3 depicts that the gait period (walking speed) showed small variations as a function of the delay conditions. We found a first maximum of the gait period at ~0.4 s, followed by a minimum at ~0.6 s, a maximum at ~0.9 s, another minimum at ~1.2 s, and a final maximum at ~1.5 s. The delay conditions with smallest gait period (or fastest walking speed) overlapped with the delay conditions for which we found that agency judgments reached maximal values (compare with Figure 2A),

comparable to those at the minimal delay (16 ms). These data show that participants' walking speed depended on the delay period, increasing after each stride and decreasing at each stride, although they were not aware of this (as revealed by questioning after the end of the experiment) and were instructed to walk at a normal habitual speed throughout the experiment.

Gait-corrected agency judgments

This makes it possible that the decrease in sine wave amplitude that we observed for agency judgments across delays is a consequence of the phase relationship between auditory delay and individual gait period. We converted the delay conditions (in s) to normalized delay conditions for each participant (the proportion of the delay with respect to the participant's gait period). The resulting data are represented in Figure 4. As can be seen in Figure 4, these gait-corrected

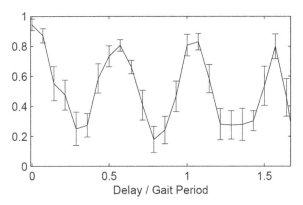

Figure 4. Gait-corrected agency judgments (mean ± 1 *SD*) across normalized delay conditions. Delay conditions (in s) were converted to normalized delay conditions for each participant (the proportion of the delay with respect to the participant's gait period). Values on the *x*-axis represent the gait-normalized delay conditions and values on the *y*-axis show agency judgments. Concerning normalized delay conditions, 0 and 1 represent time points of objective simultaneity. That is, they represent the conditions in which auditory and sensorimotor events objectively were in phase (although for the 1.0 condition, the auditory signal participants heard while simultaneously taking a step with the left foot was actually generated from the previous left footstep). The 0.5 and 1.5 points also represent points of objective simultaneity, but this time with pedal (left/right) crossover. So, in these conditions, simultaneous to taking a *left* footstep, participants heard the auditory signal associated with a previous *right* footstep. Note that the absence of agency judgment decreases around the time points of objective simultaneity.

agency data show no decrease or damping in sine wave amplitude across delays, suggesting that the decrease observed for the gait-uncorrected agency data is due to an influence of the variable gait periods between averaged individuals. Gait-corrected agency data are best fit by a sine wave function, and this function explains a significant amount of the variance (R^2 = .92, $p < .05$). Based on this model, we note that at the points of objective simultaneity, participants reliably made confirmatory agency judgments (based on the model that was the case for approximately 75% of trials). Recalibration of the observed 75% threshold with actual gait periods suggests the range of delays across which perceived simultaneity reliably was reported to be ~113 ms.

DISCUSSION

Auditory and visual effects on gait agency

Confirmatory gait agency judgments (the percentage of "yes" responses) in the present experiment decreased rapidly for delays > 120 ms and reached a first minimum at 400–500 ms. To the best of our

knowledge, agency during gait movements has not been tested using either audio-motor or visuo-motor conflicts. Our data may be compared with those described by Sato and Yasuda (2005; Experiment 2) testing agency for hand movements using finger button presses and audio-motor conflicts. These authors also observed strongly diminished agency judgments for delays > 250 ms that further decreased until 600 ms. Yet this comparison is hampered by the fact that the actions tested by Sato and Yasuda (2005) were single, goal-directed movements that were in addition associated with two different auditory action consequences (congruent and incongruent tones). This also applies to the comparison of our data with those reported by Asai & Tanno (2008), who also used finger button presses and manipulated the timing of auditory action consequences, but reported a strong decrease in such agency judgments for delays > ~100 ms. Comparison with the interesting auditory agency data by Knoblich and Repp is also difficult, as experts and naïve participants were asked to detect the moment of transition between externally and self-controlled tapping sequences (Knoblich & Repp, 2009; Repp & Knoblich, 2007). Our data manipulating performance-related auditory cues are comparable with data manipulating performance-related visual cues. Thus, Franck et al. (2001) measured agency judgments of hand actions (joystick movements) and tested different delays between a movement and a visually presented movement. They observed mostly self agency judgments for delays < 100 ms and mostly non-self agency judgments for delays > 150 ms. A comparable value of visual-motor delay has been reported by Farrer et al. (2008; Study 1; ~120 ms) and Shimada et al. (2010; ~230 ms) using similar hand movement tasks and feedback conditions. Collectively, these studies using visuo-motor and audio-motor conflict report similar values at which agency judgments decrease, ~100–200 ms. These findings are also similar to those employing *spatial* visuo-motor conflict to manipulate agency (i.e. Daprati et al., 1997; Fourneret & Jeannerod, 1998; Franck et al., 2001; Knoblich & Kircher, 2004; van den Bos & Jeannerod, 2002).

This similarity when judging temporal delays between visuo-motor and audio-motor cues is surprising considering the excellent temporal resolution of the auditory system and the relatively poor temporal resolution of the visual system. Moreover, the delay values of ~100–200 ms are far above the threshold for auditory temporal order judgments of ~20 ms (Hirsh, 1959 and Hirsh & Sherick, 1961; cited in Knoblich & Repp, 2009). We therefore suggest that this similarity reflects common, supramodal, mechanisms in the conscious action monitoring of auditory and visual action consequences. This is compatible with a comparison of the predicted and actual consequences of actions and gait (Frith et al., 2000) or the presence of a dedicated "who-system" (Georgieff & Jeannerod, 1998; Jeannerod, 2006) that is independent of the sensory modality tested. Given the relative rarity of studies testing the effects of performance-related auditory cues on agency, this has to be regarded with caution and may depend on the employed agency manipulation and task and may differ between explicit and implicit agency judgments (Repp & Knoblich, 2007).

Gait and bodily self-consciousness

Although gait has been considered initially a largely automatic action regulated mainly by subcortical control mechanisms, recent work has highlighted the influence of attentional, executive and other cognitive mechanisms (Shaw, 2002; Yogev-Seligmann, Hausdorff, & Giladi, 2008). Walking is a complex task involving the integration of locomotion, balance, and adaptation in an ever-changing environment (Armstrong, 1988; Blanc et al., 1999; Drew, Prentice, & Schepens, 2004). Moreover, the neuroscience of upright gait is hampered by two main caveats. Neuroimaging using functional magnetic resonance imaging (fMRI), magnetoencephalography (MEG), or electroencephalography (EEG) is currently not available or severely limited in walking humans (but see Fukuyama et al., 1997; Miyai et al., 2001) and humans are the only truly upright walking primates (i.e. Eccles, 1989). The neuroscience of walking is thus almost entirely based on findings in quadrupeds and behavioral work in patients with gait disorders, pointing to a distributed network including spinal cord, brainstem, basal ganglia, cerebellum, motor and posterior parietal cortex (motor cortex: Armstrong, 1988; Drew et al., 2004; spinal cord: Grillner & Wallen, 1985; Nutt, Marsden, & Thompson, 1993). The present data extend the data on cognitive gait mechanisms (Lundin-Olsson, Nyberg, & Gustafson, 1997; Shaw, 2002; Yogev-Seligmann et al., 2008) by revealing systematic conscious contributions to gait control. This is compatible with findings suggesting that human gait is a complex higher form of movement characterized by many cognitive components and likely represented at the cortical level.

In the case of conscious control for arm or hand actions awareness is lacking for the movement of a certain body part (in general the arm or the hand) of the agent, but not for the position and locomotion of the agent's entire body. We have proposed that the "I"

of conscious experience of the self – which has been linked to full-body representations (Blanke & Metzinger, 2009) – is not altered in these agency manipulations of body parts. We propose that the present procedure allowed us to manipulate the "I" of conscious experience or global bodily self, extending previous work on full-body ownership to full-body agency (Ehrsson, 2007; Kannape et al., in press; Lenggenhager et al., 2007, 2009; Petkova and Ehrsson, 2008). However despite these different functional consequences of the movement of a person's body part or a person's entire body, the present experimental data suggest that humans rely on comparable mechanisms for monitoring the action of a single body part (i.e. arm) and their entire body.

Periodic gait agency

Due to the gait cycle and continuously alternating right and left footsteps, those delays approaching the time of the subsequent actual footstep were found to increase agency judgments. Our data using gait-correction show that participants under the present experimental conditions were not aware of such moments of objective simultaneity between auditory cues and subsequent actual footsteps. This was found for all moments of objective simultaneity (with and without pedal crossover). When, simultaneous to taking a left (right) footstep, participants heard the auditory signal associated with a previous right (left) footstep, they showed no agency judgment differences. This suggests that conscious gait monitoring, in addition to mechanisms leading to gradual changes in agency judgments, also depends on periodic changes in agency judgments that depend on the participant's gait period, independent of pedal crossover. Periodic agency mechanisms have also been described in trained and naïve subjects during *rhythmic tapping movements* (Knoblich & Repp, 2009; Repp & Knoblich, 2007). We speculate that the reported independence of pedal crossover is only present in "experts" in *rhythmic stepping movements* and that there will be no comparable or less strong effects of manual crossover in tapping studies.

Our data show that it is important to include the actual movement in analysis, revealing that participants modified their walking speed depending on the tested delay condition. Moreover, we found a highly systematic influence of delay condition on walking speed, leading to an increase for auditory delays that were shorter than the first, second, and third subsequent footstep; followed by a decrease for auditory delays that were longer than the first, second, and third subsequent footstep. The automatic modulation of gait period by auditory delay is reminiscent of some cognitive effects on gait (Shaw, 2002; Yogev-Seligmann et al., 2008) as dual task performance (i.e., counting or speaking while waking) may lead to changes in walking speed that may in some subjects (especially the elderly) even lead to gait arrest and falling (Lundin-Olsson et al., 1997; Yogev-Seligmann et al., 2008). The automatic modulation of gait period is also reminiscent of spatial corrections of hand movement trajectories during hand agency judgments (Fourneret & Jeannerod, 1998; Nielsen, 1963) for which subjects are unaware.

CONCLUSION

The purpose of the present study was to investigate agency for the entire body by testing auditory action effects related to gait. For this we designed a portable stimulation and recording system in combination with performance-related auditory cues, allowing us to induce online sensorimotor conflicts and changes in agency judgments in moving agents. We show that the control of full-body locomotion and the building of a conscious experience of it are at least partially distinct brain processes. A comparison with the previous literature revealed that these delay-related agency mechanisms were similar whether auditory or visual consequences of actions were tested, compatible with supramodal, modality-independent control mechanisms. We argue that the further study of agency and ownership for a person's full body may help to refine our scientific criteria of selfhood (Blanke & Metzinger, 2009) and are of relevance for neurological conditions (Arzy, Seeck, Ortigue, Spinelli, & Blanke, 2006; Blanke et al., 2003) and psychiatric conditions (Daprati et al., 1997; Franck et al., 2001) characterized by a disturbance of selfhood.

REFERENCES

Armstrong, D. M. (1988). The supraspinal control of mammalian locomotion. *Journal of Physiology, 405*, 1–37.

Arzy, S., Seeck, M., Ortigue, S., Spinelli, L., & Blanke, O. (2006). Induction of an illusory shadow person. *Nature, 443*(7109), 287.

Asai, T., & Tanno, Y. (2008). Highly schizotypal students have a weaker sense of self-agency. *Psychiatry and Clinical Neurosciences, 62*(1), 115–119.

Aspell, J. E., Lenggenhager, B., & Blanke, O. (2009). Keeping in touch with one's self: Multisensory mechanisms of self-consciousness. *PLoS One, 4*(8), e6488.

Blanc, Y., Balmer, C., Landis, T., & Vingerhoets, F. (1999). Temporal parameters and patterns of the foot roll over during walking: Normative data for healthy adults. *Gait Posture, 10*(2), 97–108.

Blanke, O., & Metzinger, T. (2009). Full-body illusions and minimal phenomenal selfhood. *Trends in Cognitive Sciences, 13*(1), 7–13.

Blanke, O., Ortigue, S., Coeytaux, A., Martory, M. D., & Landis, T. (2003). Hearing of a presence. *Neurocase, 9*(4), 329–339.

Daprati, E., Franck, N., Georgieff, N., Proust, J., Pacherie, E., Dalery, J., et al. (1997). Looking for the agent: An investigation into consciousness of action and self-consciousness in schizophrenic patients. *Cognition, 65*(1), 71–86.

Drew, T., Prentice, S., & Schepens, B. (2004). Cortical and brainstem control of locomotion. *Progress in Brain Research, 143*, 251–261.

Eccles, J. C. (1989). Evolution of the hominid brain: Bipedality; agility. In J. C. Eccles (Ed.), *Evolution of the brain: Creation of the self* (Vol. 1, pp. 39–69). London, UK: Routledge.

Ehrsson, H. H. (2007). The experimental induction of out-of-body experiences. *Science, 317*(5841), 1048.

Farrer, C., Franck, N., Georgieff, N., Frith, C. D., Decety, J., & Jeannerod, M. (2003). Modulating the experience of agency: A positron emission tomography study. *Neuroimage, 18*(2), 324–333.

Farrer, C., Frey, S. H., Van Horn, J. D., Tunik, E., Turk, D., Inati, S., et al. (2008). The angular gyrus computes action awareness representations. *Cerebral Cortex, 18*(2), 254–261.

Fourneret, P., & Jeannerod, M. (1998). Limited conscious monitoring of motor performance in normal subjects. *Neuropsychologia, 36*(11), 1133–1140.

Franck, N., Farrer, C., Georgieff, N., Marie-Cardine, M., Dalery, J., d'Amato, T., et al. (2001). Defective recognition of one's own actions in patients with schizophrenia. *American Journal of Psychiatry, 158*(3), 454–459.

Frith, C. D., Blakemore, S. J., & Wolpert, D. M. (2000). Abnormalities in the awareness and control of action. *Philosophical Transactions of the Royal Society of London, Series B: Biological Sciences, 355*(1404), 1771–1788.

Fukuyama, H., Ouchi, Y., Matsuzaki, S., Nagahama, Y., Yamauchi, H., Ogawa, M., et al. (1997). Brain functional activity during gait in normal subjects: A SPECT study. *Neuroscience Letters, 228*(3), 183–186.

Georgieff, N., & Jeannerod, M. (1998). Beyond consciousness of external reality: A "who" system for consciousness of action and self-consciousness. *Consciousness and Cognition, 7*(3), 465–477.

Grillner, S., & Wallen, P. (1985). Central pattern generators for locomotion, with special reference to vertebrates. *Annual Review of Neuroscience, 8*, 233–261.

Hirsh, I. J. (1959). Auditory perception of temporal order. *Journal of the Acoustical Society of America, 31*, 759–767.

Hirsh, I. J., & Sherrick, C. E. (1961). Perceived order in different sense modalities. *Journal of Experimental Psychology, 62*, 423–432.

Jeannerod, M. (2006). The origin of voluntary action: History of a physiological concept. *Comptes Rendus Biologies, 329*(5–6), 354–362.

Jeannerod, M. (2007). Being oneself. *Journal of Physiology (Paris), 101*(4–6), 161–168.

Kannape, O. A., Schwabe, L., Tadi, T., & Blanke, O. (in press). The limits of agency in walking humans. *Neuropsychologia*.

Knoblich, G., & Kircher, T. T. (2004). Deceiving oneself about being in control: Conscious detection of changes in visuomotor coupling. *Journal of Experimental Psychology: Human Perception and Performance*, *30*(4), 657–666.

Knoblich, G., & Repp, B. H. (2009). Inferring agency from sound. *Cognition*, *111*(2), 248–262.

Lenggenhager, B., Mouthon, M., & Blanke, O. (2009). Spatial aspects of bodily self-consciousness. *Consciousness and Cognition*, *18*(1), 110–117.

Lenggenhager, B., Tadi, T., Metzinger, T., & Blanke, O. (2007). Video ergo sum: Manipulating bodily self-consciousness. *Science*, *317*(5841), 1096–1099.

Lundin-Olsson, L., Nyberg, L., & Gustafson, Y. (1997). "Stops walking when talking" as a predictor of falls in elderly people. *Lancet*, *349*(9052), 617.

Macellari, V., Giacomozzi, C., & Saggini, R. (1999). Spatial–temporal parameters of gait: Reference data and a statistical method for normality assessment. *Gait Posture*, *10*(2), 171–181.

Miyai, I., Tanabe, H. C., Sase, I., Eda, H., Oda, I., Konishi, I., et al. (2001). Cortical mapping of gait in humans: A near-infrared spectroscopic topography study. *NeuroImage*, *14*(5), 1186–1192.

Murray, M. P. (1967). Gait as a total pattern of movement. *American Journal of Physical Medicine, 46*, 290–333.

Nielsen, T. (1963). Volition: A new experimental approach. *Scandinavian Journal of Psychology*, *4*(4), 225–230.

Nutt, J. G., Marsden, C. D., & Thompson, P. D. (1993). Human walking and higher-level gait disorders, particularly in the elderly. *Neurology*, *43*(2), 268–279.

Petkova, V. I., & Ehrsson, H. H. (2008). If I were you: Perceptual illusion of body swapping. *PLoS One*, *3*(12), e3832.

Repp, B. H., & Knoblich, G. (2007). Toward a psychophysics of agency: Detecting gain and loss of control over auditory action effects. *Journal of Experimental Psychology: Human Perception and Performance*, *33*(2), 469–482.

Sato, A., & Yasuda, A. (2005). Illusion of sense of self-agency: Discrepancy between the predicted and actual sensory consequences of actions modulates the sense of self-agency, but not the sense of self-ownership. *Cognition*, *94*(3), 241–255.

Shaw, F. E. (2002). Falls in cognitive impairment and dementia. *Clinics in Geriatric Medicine*, *18*(2), 159–173.

Shimada, S., Qi, Y., & Hiraki, K. (2010). Detection of visual feedback delay in active and passive self-body movements. *Experimental Brain Research*, *201*, 359–364.

Tsakiris, M., Haggard, P., Franck, N., Mainy, N., & Sirigu, A. (2005). A specific role for efferent information in self-recognition. *Cognition*, *96*(3), 215–231.

van den Bos, E., & Jeannerod, M. (2002). Sense of body and sense of action both contribute to self-recognition. *Cognition*, *85*(2), 177–187.

Yogev-Seligmann, G., Hausdorff, J. M., & Giladi, N. (2008). The role of executive function and attention in gait. *Movement Disorders*, *23*(3), 329–342; quiz 472.

COGNITIVE NEUROSCIENCE, 2010, 1 (3), 193–203

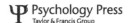

Disorders of consciousness: Moving from passive to resting state and active paradigms

M. A. Bruno[1], A. Soddu[1], A. Demertzi[1], S. Laureys[1,2], O. Gosseries[1], C. Schnakers[1], M. Boly[1], Q. Noirhomme[1], M. Thonnard[1], C. Chatelle[1], and A. Vanhaudenhuyse[1]

[1]University and University Hospital of Liège, Liège, Belgium
[2]University Hospital of Liège, Liège, Belgium

Following coma, some patients will recover wakefulness without signs of consciousness (i.e., vegetative state) or may show nonreflexive movements but with no ability for functional communication (i.e., minimally conscious state). Currently, there remains a high rate of misdiagnosis of the vegetative state. The increasing use of fMRI and EEG tools permits the clinical characterization of these patients to be improved. We first discuss "resting metabolism" and "passive activation" paradigms, used in neuroimaging and evoked potential studies, which merely identify neural activation reflecting "automatic" processing—that is, occurring without the patient's willful intervention. Secondly, we present an alternative approach consisting of instructing subjects to imagine well-defined sensory-motor or cognitive-mental actions. This strategy reflects volitional neural activation and, hence, witnesses awareness. Finally, we present results on blood-oxgen-level-dependent "default mode network"/resting state studies that might be a promising tool in the diagnosis of these challenging patients.

Keywords: Consciousness; Vegetative state; Resting state; fMRI; Active paradigms.

INTRODUCTION

Following severe brain damage, patients classically evolve through different clinical stages before recovering consciousness. Coma is usually a transient condition that will last no longer than a few days or weeks and is defined as an "unarousable unresponsiveness" (Posner, Saper, Schiff, & Plum, 2007). After some days or weeks, comatose patients may open their eyes. When this return to "wakefulness" is accompanied by reflexive motor activity only, the condition is called vegetative state (VS; Multi-Society Task Force on PVS, 1994). Many patients in VS regain consciousness in the first month after brain injury. However, if patients show no signs of awareness 1 year after a traumatic brain injury or 3 months after anoxic accident, the chances of recovering functional communication are close to zero and the patient is considered in a permanent VS (Multi-Society Task Force on PVS, 1994). Those who recover from VS typically progress to a minimally conscious state (MCS), which is characterized by minimal but definite behavioral evidence of awareness of self and/or the environment (Giacino et al., 2002). Like the VS, the MCS may be chronic and sometimes permanent. However, at present, no time-windows for "permanent

Correspondence should be addressed to: S. Laureys, Coma Science Group, Cyclotron Research Center, University and University Hospital of Liège, Sart-Tilman B30, Liège, Belgium. E-mail: steven.laureys@ulg.ac.be

M. A. Bruno and A. Soddu contributed equally to the manuscript.

SL is Senior Research Associate; AS, MB and QN are Post-doctoral Fellows; and MAB is Research Fellow at the Fonds de la Recherche Scientifique (FRS). This research was supported by the Fonds de la Recherche Scientifique (FRS), the European Commission (DISCOS, Mindbridge, DECODER, & CATIA), Concerted Research Action (ARC-06/11-340), The McDonnell Foundation, and The Mind Science Foundation.

        DOI: 10.1080/17588928.2010.485677

MCS" have been agreed upon. Reliable and consistent interactive communication and/or functional use of objects indicate the next boundary in the course of recovery: emergence from MCS (Giacino et al., 2002). Finally, some patients may awake from their coma but be unable to move or speak: the only way they have to express their consciousness is by small eye movements. This state is called locked-in syndrome (LIS) and is characterized by quadriplegia and anarthria with general preservation of cognition (Schnakers et al., 2008a) and an elementary mode of communication that uses vertical or lateral eye movements (American Congress of Rehabilitation Medicine, 1995).

Behavioral assessment is one of the main methods used to detect awareness in severely brain injured patients recovering from coma. Clinical practice shows that recognizing unambiguous signs of conscious perception in such patients can be very challenging. Indeed, behavioral assessment is complicated by the presence of motor/language impairment, tracheotomy, fluctuating arousal level or ambiguous and rapidly habituating responses, which may partly explain the high frequency of misdiagnosis (up to 43%) in these patients (Schnakers et al., 2009b). This underestimation of the patient's level of consciousness can also be explained by other factors such as poor expertise in behavioral assessment or the use of insensitive behavioral assessment tools (e.g., using a mirror when evaluating visual pursuit, behavior that is considered as one of the first clinical signs differentiating MCS from VS; Vanhaudenhuyse, Schnakers, Bredart, and Laureys, 2008b). LIS may also be misdiagnosed as coma or VS possibly due to the rarity of this syndrome, the difficulty of recognizing unambiguous signs of consciousness, the fluctuation of vigilance in the acute setting or by additional cognitive or sensory deficits, such as deafness (Bruno et al., 2009; Schnakers et al., 2008a; Smart et al., 2008). Misdiagnosis can lead to grave consequences, especially in end-of-life decision-making. Contrary to patients in VS, those in MCS retain some capacity for cognitive processing (see next section). Moreover, the prognosis of MCS patients is significantly more favorable relative to those in VS (Giacino, 2004a). Clinical attitudes and decisions about end-of-life and pain, therefore, are likely to be influenced by whether one is diagnosed as a VS or a MCS patient (Demertzi et al., 2009). New technical methods offer specific assessment procedures and the possibility to determine objectively whether an unresponsive patient is aware without explicit verbal or motor response. We here review neuroimaging studies using positron emission tomography (PET), functional magnetic resonance imaging (fMRI), and electrophysiological techniques for the assessment of patients with disorders of consciousness, and present new data on resting state brain activity in VS patients.

GLOBAL RESTING METABOLISM

Using PET, Levy et al. (1987) first showed that patients in a VS suffer from a massive cerebral metabolic reduction, estimated to be 40–50% of normal values. These results have been repeatedly confirmed by others (Laureys et al., 1999a; Rudolf, Ghaemi, Haupt, Szelies, & Heiss, 1999; Tommasino, Grana, Lucignani, Torri, & Fazio, 1995) in VS of different etiologies and duration. In patients with an LIS, overall supratentorial cerebral metabolism has been shown to be preserved partially or fully (Levy et al., 1987), whereas in comatose patients a decrease of 45% in cerebral metabolism has been observed (Laureys et al., 2001). However, a global depression of cerebral metabolism is not specific to VS or coma only. In slow wave sleep, overall brain metabolism also decreases approximately to 40% of normal waking values while in REM-sleep metabolism returns to normal values (Maquet et al., 1990). Another example of transient metabolic depression is observed during general anesthesia, which is characterized by comparable reduction in cortical metabolism to that observed in VS (Alkire et al., 1999). However, we have shown that the relationship between global levels of brain function and the presence or absence of awareness is not absolute; rather, some areas in the brain seem more important than others for the emergence of awareness (Laureys, Faymonville, Moonen, Luxen, & Maquet, 2000c; Laureys et al., 1999a). Indeed, VS patients who subsequently recovered consciousness did not show substantial changes in global metabolic rates for glucose metabolism (Laureys, Lemaire, Maquet, Phillips, & Franck, 1999b), and some awake healthy volunteers have global brain metabolism values comparable to those observed in some patients in a VS (Laureys, 2005). In VS, a dysfunction was found not in the whole brain but in a wide frontoparietal network encompassing polymodal associative cortices: bilateral lateral frontal regions; parieto-temporal and posterior parietal areas; mesiofrontal, posterior cingulated, and precuneal cortices (Laureys et al., 1999a, 2004a). Posterior cingulate and adjacent precuneal cortices were reported to differentiate MCS from VS patients (Laureys, Owen, & Schiff, 2004a; intermediate metabolism in MCS, higher than in VS but lower than in conscious controls). More recently, a study on patients in the chronic stage of traumatic diffuse brain injury showed a bilateral hypometabolism in the medial prefrontal regions, the medial frontobasal regions, the cingulate gyrus, and the thalamus. At the group level, these regions were more hypometabolic in VS than MCS patients, but MCS still showed less activation than patients who have emerged from this state and who recovered the ability to communicate (Nakayama, Okumura, Shinoda, Nakashima, & Iwama, 2006). Specifically, awareness seems related not exclusively to the activity in the frontoparietal network but, as importantly, to the functional connectivity within this network and the thalami. Indeed, long-range cortico-cortical and cortico-thalamo-cortical "functional disconnections" could be identified in the VS (Laureys et al., 1999a, 2000b).

PASSIVE PARADIGMS: CEREBRAL ACTIVATION FROM EXTERNAL STIMULATION

Functional neuroimaging

Somatosensory, auditory, and visual perceptions are conscious experiences; thus, the wakeful unconsciousness of VS patients, by definition, precludes these experiences. However, the absence of a behavioral response cannot be taken as an absolute proof of the absence of consciousness. Several fMRI activation studies in VS (Coleman et al., 2007; Di et al., 2007; Fernandez-Espejo et al., 2008; Moritz et al., 2001; Staffen, Kronbichler, Aichhorn, Mair, and Ladurner, 2006) have confirmed previous PET studies showing preserved activation of "lower level" primary sensory cortices which are disconnected from "higher order" associative cortical networks employing auditory (Boly et al., 2004; Laureys et al., 2000a; Owen et al., 2002), somatosensory (Boly et al., 2008a), or visual (Menon et al., 1998; Owen et al., 2002) stimulations. Similar studies in PET reported that MCS patients showed a more widespread activation than VS, with a cortico-cortical functional connectivity more efficient in MCS compared to VS (Boly et al., 2004). These results confirmed that MCS patients may experience pain and hence should systematically receive appropriate analgesic treatment. Moreover, stimuli with emotional valence (baby cries and own name) were shown to induce a much more widespread activation than did meaningless noise in the MCS (Laureys et al., 2004b). Stimuli

with emotional valence (the voice of the patient's mother compared with an unfamiliar voice) were also shown to activate amygdala in a traumatic MCS patient (Bekinschtein et al., 2004). Similarly, Schiff et al. (2005), in two MCS patients, showed selective activation in components of the cortical language networks during presentation of narratives read by a familiar voice and containing personally meaningful content. Such context-dependent higher-order auditory processing shows that content does matter when talking to MCS patients. Finally, some exceptional VS patients may also show higher atypical level of cortical activation in response to auditory stimulations, and this was proposed to be a surrogate marker of good prognosis (Di et al., 2007).

Event-related potentials

Event-related potentials (ERPs) have been used for a long time to assess comatose patients. Early components of these potentials arising within 100 ms are known to persist even in unconscious states. The later components of exogenous potentials and other so-called "endogenous" ERP components (e.g., P300) are more reliably related to the (unconscious or conscious) cognitive processing of the information, and less frequently observed in disorders of consciousness (Vanhaudenhuyse, Laureys, & Perrin, 2008a). In several studies, ERPs were used to evaluate the integrity of detection of noncommunicative patients' own name, in order to assess the possible preservation of residual linguistic and self-processing in these patients. Differential P300 wave to the own name (as compared to other names) was observed in LIS patients (Perrin et al., 2006), which is not surprising since their cognitive functions and their linguistic comprehension remain preserved (Onofrj, Thomas, Paci, Scesi, and Tombari, 1997; Schnakers et al., 2008a). In MCS, results suggested that the auditory system was relatively preserved in response to passive tones and language stimulation, implying that these patients are able to detect salient words (Perrin et al., 2006). Most surprisingly, some VS patients emitted a differential P300, although delayed as compared to age-matched controls (Perrin et al., 2006). These last results showed that the P300 wave resulting from a passive paradigm is not useful to successfully distinguish unconscious from conscious patients. However, in most studies, the presence of a P300 correlated with favorable outcome in comatose patients (Vanhaudenhuyse et al., 2008a).

ACTIVE PARADIGMS: COMMAND FOLLOWING IN NONCOMMUNICATIVE PATIENTS

Functional neuroimaging

In the absence of a full understanding of the neural correlates of consciousness, even a near-normal activation in response to passive stimulations cannot be considered a proof of the presence of awareness. Instead, all that can be inferred is that a specific brain region is, to some degree, still able to perceive and process relevant sensory stimuli. The question that arises is how we can disentangle automatic from voluntary conscious brain activation. In 2006, Owen et al. addressed this concern by applying an fMRI paradigm in a traumatic VS patient who was asked to perform two mental imagery tasks ("Imagine playing tennis" and "Imaging visiting your house"). Activation was observed in the supplementary motor area after the patient was asked to imagine playing tennis, and in parahippocampal gyrus when she was asked to imagine visiting her house. Similar activation patterns were seen in healthy volunteers (Boly et al., 2007a). Importantly, because the only difference between the conditions that elicited task-specific activation was in the instruction given at the beginning of each scanning session, the activation observed can only reflect the intentions of the patient, rather than some altered property of the outside world. Some could argue that the words "tennis" and "house" may have automatically triggered the patterns of activation observed in target brain areas in this patient in the absence of conscious awareness. Although it is well documented that these words elicit wholly automatic neural responses in the absence of conscious awareness (Hauk, Johnsrude, & Pulvermuller, 2004), such responses typically last for a few seconds and occur in regions of the brain that are associated with word processing. In this patient, the observed activity persisted for the full 30 s of each imagery task and persisted until the patient was cued with another stimulus indicating that she should rest. Thus, such responses cannot be explained in terms of automatic brain processes (Soddu et al., 2009). In addition, the responses in the patient were observed not in brain regions that are known to be involved in word processing but, rather, in regions that are known to be involved in the two imagery tasks that she was asked to carry out (Owen et al., 2007). In this sense, the decision to "imagine playing tennis" rather than simply "rest" is an act of willed intention and, therefore, clear evidence for response to command awareness. The results of this study should not be misinterpreted as evidence that all patients in VS may be conscious. Interestingly, when re-examined six months later, the patient showed inconsistent signs of consciousness. The most likely explanation of these results is that the patient was already beginning the transition to the MCS at the time of the experiment. This study also highlights the importance of fMRI as a potentially good marker for both diagnosis and prognosis (Di et al., 2007).

ERPs

The use of a passive ERP paradigm is not sufficient to reliably disentangle VS from MCS. Indeed, even if passive ERP paradigms can highlight ongoing brain processing for a given stimulus input, they do not differentiate between automatic and voluntary cognitive processes and, therefore, between unconscious and conscious patients. For this reason, an active ERPs paradigm was developed, where the participant is instructed to voluntarily direct attention to a target stimulus and to ignore other stimuli (Figure 1) (Schnakers et al., 2008b, 2009a). Group as well as individual results showed that a larger P300 response to the own name was observed in MCS patients in active condition as compared to passive listening. This P300 amplitude was otherwise equivalent to that observed in controls, while no task-related P300 changes in VS patients were observed (Schnakers et al., 2008b). This suggests that MCS patients were able to voluntarily focus their attention on the target as a function of task requirements. This active ERPs paradigm permitted to detect consciousness in a total LIS patient (i.e., characterized by complete immobility including all eye movements; Bauer, Gerstenbrand, & Rumpl, 1979) that behaviorally would be diagnosed as comatose (Schnakers et al., 2009a). In a similar way, conscious processing was detected in three out of four MCS patients who were instructed to actively count the number of deviant trials in series of sound, while no response were recorded in VS patients (Bekinschtein et al., 2009).

RESTING STATE fMRI PARADIGM: A NEW TOOL TO CATEGORIZE DISORDERS OF CONSCIOUSNESS PATIENTS?

Resting state fMRI acquisitions are easy to perform and could have a potentially broader and faster translation into clinical practice. Recent studies on spontaneous fluctuations in the fMRI blood-oxygen-level-dependent (BOLD) signal recorded in "resting" awake healthy subjects showed the presence of coherent fluctuations

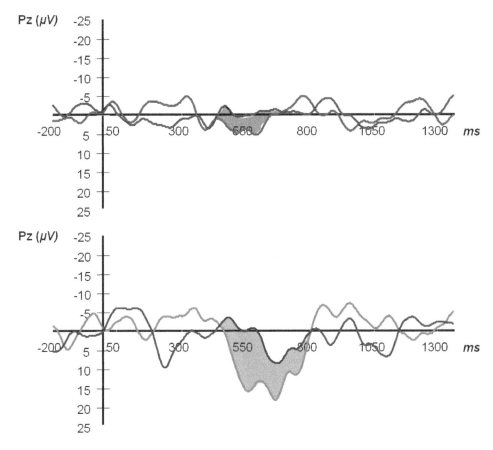

Figure 1. Illustration of active and passive event-related potential paradigms in a 22-year-old locked-in syndrome patient. Upper panel depicts the patient's response to an unfamiliar name (listened nontarget unfamiliar name, in blue) as compared with active condition (responded to target unfamiliar name, in red). Lower panel depicts the patient's own-name response in the passive condition (listened nontarget own name, in green) and in the active condition (responded to target own name, in pink). (Adapted from data in Schnakers et al., 2009a.)

among functionally defined neuroanatomical networks (Boly et al., 2008b; Raichle, 2006). The concept of a "default mode network" describes a set of brain areas exhibiting task-induced deactivations, encompassing precuneus, posterior parietal lobe, and medial prefrontal cortex, which are more active at rest than when we are involved in attention-demanding cognitive tasks (Raichle et al., 2001). The clinical interest of default network MRI studies is that they allow the investigation of higher order cognitive networks without requiring patients' active participation, particularly important in VS and MCS. We here compared default mode network in VS and brain-dead patients with healthy volunteers.

Patients

We first created a healthy control template of the default mode network (DMN) by analyzing 11 volunteers (age range 21–60 years; 4 women). Secondly, we compared 4 brain injured patients (3 vegetative:

1 ischemia, 1 encephalopathy, 1 traumatic; and 1 hemorrhagic brain-dead patients (published in Boly et al., 2009), age range 27–77 years, all men) to 9 other healthy volunteers (age range 29–65 years, 5 women). In patients, clinical examination was repeatedly performed using standardized scales (the Coma Recovery Scale Revised, Giacino Kalmar, & Whyte, 2004b; and the Glasgow Liege scale, Born, 1988) on the day of scanning, and in the week before and the week after. Patients were scanned in an unsedated condition. The study was approved by the Ethics Committee of the Medical School of the University of Liège. Informed consent to participate to the study was obtained from the subjects themselves in the case of healthy subjects, and from the legal surrogates of the patients.

Data acquisition and analysis

In all participants, resting state BOLD data were acquired on a 1.5 T MR scanner (Symphony Tim,

Siemens, Germany) with a gradient echo-planar sequence using axial slice orientation (36 slices; voxel size = 3.75 × 3.75 × 3.6 mm³; matrix size = 64 × 64 × 36; repetition time = 3000 ms; echo time = 30 ms; flip angle = 90°; field of view = 240 mm). A protocol of 200 scans with a duration of 10 min was performed. fMRI data were preprocessed using BrainVoyager software (R. Goebel, Brain Innovation, Maastricht, The Netherlands). Preprocessing of functional scans included 3D motion correction, linear trend removal, slice scan time correction, and filtering out of low frequencies of up to 0.005 Hz. The data were spatially smoothed with a Gaussian filter of full width at half maximum value of 8 mm. Independent component analysis (ICA) was performed with BrainVoyager using 30 components (Ylipaavalniemi & Vigario, 2008). An average template DMN map was calculated on 11 healthy subjects. We performed, as implemented in BrainVoyager (self-organizing ICA; Esposito et al., 2005), a spatial similarity test on single subjects independent components (IC) and we averaged the maps of the ICs belonging to the same cluster. The cluster corresponding to the DMN was selected by visual inspection. In order to select the DMN for each subject and patient, we run self-organizing ICA with the average ICs coming from the 11 healthy subjects template and we picked the component clustering with the DMN template (spatial similarity test). Self-organizing ICA could assign a similarity value that indicates how well the selected IC fitted the average DMN map based on the 11 subjects (the similarity test was based only on spatial properties without excluding components with a power spectrum dominated by high frequency). The same protocol of 200 scans was acquired on a spherical phantom. Finally, a two-tailed unequal variance Student t-test compared the spatial similarity values of the selected DMN map in healthy controls compared to VS patients. Three motion indices were introduced describing the motion of patients compared to healthy controls: the mean frequency calculated as the mean of the frequency of each motion curve (three translations and three rotations), the mean over time of the full displacement during the acquisition calculated as the square root of the sum of the squares of the six motion parameters, and the mean over time of the displacement speed during the acquisition calculated as the square root of the sum of the squares of the six motion parameters variation over one time unit (repetition time).

RESULTS

A healthy controls mean spatial map identified the DMN pattern (Figure 2a) showing full overlap with the template identified by the black and white contour. The principal brain regions characterizing the DMN, posterior cingulate/precuneal, mesiofrontal and posterior parietal cortices, were detected. Compared to controls, VS patients showed a significant lower spatial similarity (0.47 ± 0.08, range 0.33–0.55 vs. 0.23 ± 0.12, range 0.12–0.35; $p = .05$). The brain-dead patient (and the phantom) also showed lower spatial similarity compared to controls with values in the VS range. Finally, none of the VS patients showed a DMN with a spatial pattern comparable with healthy controls (Figure 2b–d), as assessed by visual inspection, even if in the case of V1 the spatial similarity had a value consistent with healthy subjects (Table 1). Rapid transient "clonic" motions (i.e., spasmodic changes of muscular contraction) were observed for VS1 and VS3 as also confirmed by spatial maps' periphery patterns. The brain-dead patient didn't show any significant spatial pattern, confirming results of Boly and collaborators (2009).

DISCUSSION

After creating a template of the DMN, we identified the DMN at the single subject level, in a user-independent manner. In healthy volunteers, the identified component showed the typical spatial pattern of the DMN (Beckmann, DeLuca, Devlin, & Smith, 2005). In VS patients, as well as in the brain-dead, we failed to show any consistent pattern of DMN, even if single subjects brain activation maps showed residual connectivity in VS. In our view, these residual connectivity patterns are not reflecting residual DMN neuronal activity but could be explained by movement artifacts. Cardiorespiratory and vascular effects could also be a source of artifactual connectivity. BOLD signal changes within regions of the DMN have been found to be reduced after correcting for cardiorespiratory effects (van Buuren et al., 2009). In addition, it should be stressed that altered blood flow and physiology, such as neurovascular decoupling, could also effect the BOLD signal detected in fMRI (Logothetis, Pauls, Augath, Trinath, & Oeltermann, 2001), especially in the context of traumatic and ischemic encephalopathy. This of course cannot exclude the absence of neural activity in these patients (Boly et al., 2009; van Buuren et al., 2009). Boly and collaborators (2009) showed that correlations with posterior cingulate cortex are reduced in one VS as compared to age-matched controls, and argued that this reduced functional connectivity within the DMN in VS was in line with the hypothesis that spontaneous fMRI signal changes may be partly related to ongoing interoceptive states of mind or conscious thoughts.

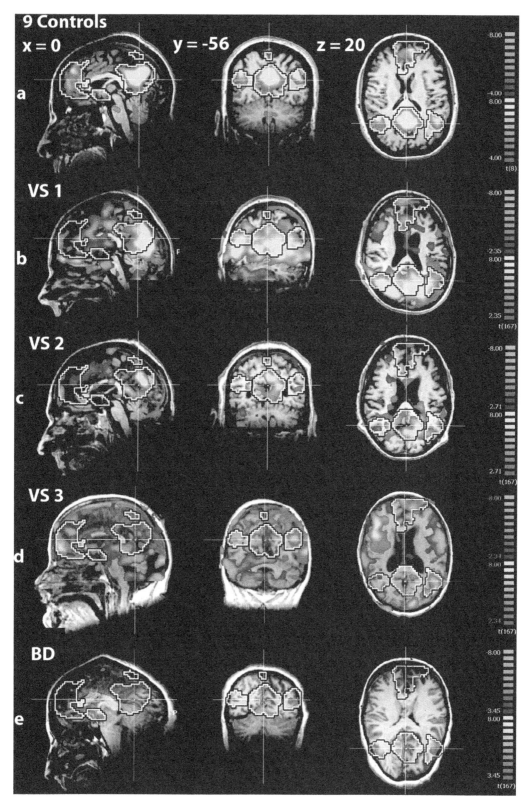

Figure 2. Resting state EPI–BOLD acquisition in (a) 9 healthy volunteers, with a spatial map obtained by running a random effect group GLM analysis using as predictors the time courses of the ICs selected as default mode (thresholded at false discovery rate corrected $p < .05$) within default mode mask obtained from an independent dataset (the 11 healthy controls for the independent study) shown as black and white contour volume of interest. (b–d) 3 vegetative patients and (e) 1 brain death with spatial maps obtained by running a GLM analysis using as predictor the time courses of the ICs selected as default mode (thresholded at false discovery rate corrected $p < .05$).

TABLE 1

Spatial and temporal properties of the independent component selected as default mode (DM) from resting state fMRI in healthy controls, vegetative (VS) and one brain death (BD) patients (phantom is added for comparison)

	Spatial similarity	Normalized variation of spatial similarity	DM time course mean frequency (Hz)	Motion curves mean frequency (Hz)	Mean displacement	Mean speed
Control 1	0.33	1.5	0.04	0.03	0.6	0.04
Control 2	0.51	2.9	0.05	0.06	0.3	0.14
Control 3	0.37	1.8	0.05	0.03	1.7	0.09
Control 4	0.50	2.8	0.05	0.06	0.4	0.14
Control 5	0.54	3.2	0.07	0.05	0.8	0.25
Control 6	0.39	2	0.04	0.04	0.3	0.02
Control 7	0.55	3.2	0.05	0.03	3.0	0.09
Control 8	0.54	3.2	0.05	0.04	0.3	0.03
Control 9	0.47	2.6	0.04	0.03	0.8	0.05
VS 1	0.35	1.7	0.04	0.05	1.1	0.14
VS 2	0.23	0.8	0.07	0.05	0.2	0.28
VS 3	0.12	−0.1	0.04	0.03	5.9	0.50
BD	0.13	0	0.07	0.06	0.1	0.27
Phantom	0.12	−0.1	0.06	0.03	0.1	0.24

Notes: Spatial similarity with an average-template-default mode based on an independent 11 healthy controls data set, normalized variation of spatial similarity (Sim) respect to the brain death (BD) (Sim_subject-Sim_BD)/Sim_BD), and default mode time course mean frequency. Motion properties (means calculated from the six motion curves). Rapid transient "clonic" motions were observed for two of the three patients (VS1 and VS3) as also confirmed by spatial maps periphery patterns, while a drift (i.e., a displacement with low speed) is observed for control 3 and 7.

Motion, pulse, and respiratory artifacts remain a very important problem to tackle to be able to properly assess patients with disorders of consciousness. If recording heart and respiratory rates during the acquisition can help in disentangling neuronal activity from pulse and respiratory artifacts (Gray et al., 2009), further investigations are needed to disentangle effects of motion and neuronal activity in the BOLD signal. In the brain-dead patient, who fulfilled all the standard clinical criteria for brain death (previously published in Boly et al., 2009), fMRI results did not show any long-range significant functional connectivity, confirming that the origin of the BOLD signal was fully due to motion artifacts as expected due to the complete absence of neuronal activity in brain death. Finally, future work should study the relationship between structural and functional connectivity changes in noncommunicative brain-damaged patients. Functional connectivity in DMN has indeed been related to underlying structural anatomy (Greicius, Supekar, Menon, & Dougherty, 2009). Further multimodal studies should combine resting state fMRI data with structural MRI (e.g. diffusion tensor imaging data).

CONCLUSION

Patients with disorders of consciousness represent a major clinical problem in terms of clinical assessment, treatment, and daily management. The nonresponsiveness of such patients implies that they can only be diagnosed by means of exclusion criteria such as "no goal-directed eye movements" or "no execution of commands." Those behavioral signs may be very slight and short-term and performance may fluctuate. Integration of neuroimaging and ERPs techniques should improve our ability to disentangle diagnostic and prognostic differences on the basis of underlying mechanisms and better guide our clinical therapeutic options in these challenging patients. A step-by-step approach combining multimodal assessment techniques (i.e., PET scan, fMRI, and ERPs) seems to be appropriate to detect signs of consciousness (Figure 3). Resting metabolism and passive paradigms studies increased our understanding of residual cerebral processing of VS and MSC patients. Active paradigms seem to provide an objective valuable additional diagnostic tool in cases of patients with atypical activation, leading to persisting doubts in clinical diagnosis. Negative results, however, must be cautiously interpreted in case of patients with severely altered level of vigilance, which could present only transient activity in response to the presentation of instructions. Passive and active fMRI paradigms were also shown to potentially be a useful tool to predict possibility of recovery.

Finally, results on the DMN seem promising to distinguish patients with disorders of consciousness by evaluating preservation of connectivity in this resting state network. However, future studies are needed to give a full characterization of default mode connectivity in VS and MCS patients and its potential use in outcome prediction. Future clinical studies should also perform

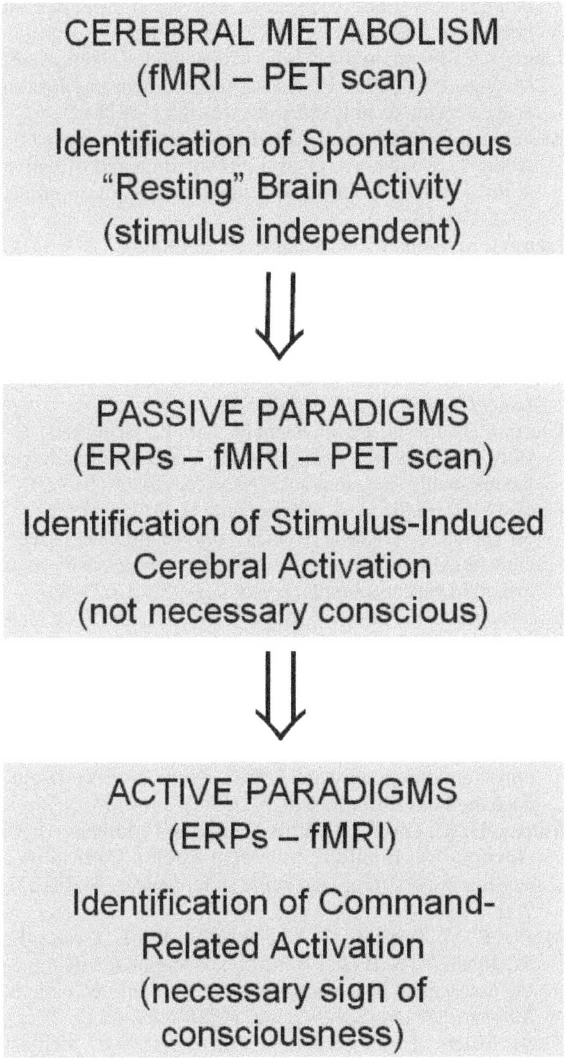

CEREBRAL METABOLISM
(fMRI – PET scan)

Identification of Spontaneous "Resting" Brain Activity (stimulus independent)

PASSIVE PARADIGMS
(ERPs – fMRI – PET scan)

Identification of Stimulus-Induced Cerebral Activation (not necessary conscious)

ACTIVE PARADIGMS
(ERPs – fMRI)

Identification of Command-Related Activation (necessary sign of consciousness)

Figure 3. Illustration of a step-by-step approach combining PET scan, ERPs, and fMRI techniques which might reveal signs of consciousness that are unattainable by bedside clinical assessment of disorders of consciousness patients. These high-tech devices might permit some of these patients to show their consciousness by following commands (e.g., to imagine playing tennis, respond to own name) via nonmotor pathways.

heart rate and respiratory rate recording and simultaneous real movement monitoring, which would improve default mode analyses.

REFERENCES

Alkire, M. T., Pomfrett, C. J., Haier, R. J., Gianzero, M. V., Chan, C. M., Jacobsen, B. P., et al. (1999). Functional brain imaging during anesthesia in humans: Effects of halothane on global and regional cerebral glucose metabolism. *Anesthesiology, 90*(3), 701–709.

American Congress of Rehabilitation Medicine (1995). Recommendations for use of uniform nomenclature pertinent to patients with severe alterations of consciousness. *Archives of Physical Medicine and Rehabilation, 76,* 205–209.

Bauer, G., Gerstenbrand, F., & Rumpl, E. (1979). Varieties of the locked-in syndrome. *Journal of Neurology, 221*(2), 77–91.

Beckmann, C. F., DeLuca, M., Devlin, J. T., & Smith, S. M., (2005). Investigations into resting-state connectivity using independent component analysis. *Philosophical Transactions of the Royal Society of London B: Biological Sciences, 360,* 1001–1013.

Bekinschtein, T., Niklison, J., Sigman, L., Manes, F., Leiguarda, R., Armony, J., et al. (2004). Emotion processing in the minimally conscious state. *Journal of Neurology, Neurosurgery & Psychiatry, 75*(5), 788.

Bekinschtein, T., Dehaene, S., Rohaut, B., Tadel, F., Cohen, L., & Naccache, L. (2009). Neural signature of the conscious processing of auditory regularities. *Proceedings of the National Academy of Sciences of the United States of America, 106*(5), 1672–1677.

Boly, M., Coleman, M. R., Davis, M. H., Hampshire, A., Bor, D., Moonen, G., et al. (2007a). When thoughts become action: An fMRI paradigm to study volitional brain activity in non-communicative brain injured patients. *NeuroImage, 36*(3), 979–992.

Boly, M., Faymonville, M. E., Peigneux, P., Lambermont, B., Damas, P., Del Fiore, G., et al. (2004). Auditory processing in severely brain injured patients: Differences between the minimally conscious state and the persistent vegetative state. *Archives of Neurology, 61*(2), 233–238.

Boly, M., Faymonville, M. E., Schnakers, C., Peigneux, P., Lambermont, B., Phillips, C., et al. (2008a). Perception of pain in the minimally conscious state with PET activation: An observational study. *Lancet Neurology, 7*(11), 1013–1020.

Boly, M., Phillips, C., Tshibanda, L., Vanhaudenhuyse, A., Schabus, M., Dang-Vu, T. T., et al. (2008b). Intrinsic brain activity in altered states of consciousness: How conscious is the default mode of brain function? *Annals of the New York Academy of Sciences, 1129,* 119–129.

Boly, M., Tshibanda, L., Vanhaudenhuyse, A., Noirhomme, Q., Schnakers, C., Ledoux, D., et al. (2009). Functional connectivity in the default network during resting state is preserved in a vegetative but not in a brain dead patient. *Human Brain Mapping, 30*(8), 2393–2400.

Born, J. D. (1988). The Glasgow-Liège Scale: Prognostic value and evaluation of motor response and brain stem reflexes after severe head injury. *Acta Neurochirurgica, 95,* 49–52.

Bruno, M. A., Schnakers, C., Damas, F., Pellas, F., Lutte, I., Bernheim, J., et al. (2009). Locked-in syndrome in children: Report of five cases and review of the literature. *Pediatric Neurology, 41*(4), 237–246.

Coleman, M. R., Rodd, J. M., Davis, M. H., Johnsrude, I. S., Menon, D. K., Pickard, J. D., et al. (2007). Do vegetative patients retain aspects of language comprehension? Evidence from fMRI. *Brain, 130*(10), 2494–2507.

Dehaene, S., & Naccache, L. (2001). Towards a cognitive neuroscience of consciousness: Basic evidence and a workspace framework. *Cognition, 79*(1–2), 1–37.

Demertzi, A., Schnakers, C., Ledoux, D., Chatelle, C., Bruno, M. A., Vanhaudenhuyse, A., et al. (2009). Different beliefs about pain perception in the vegetative and minimally conscious states: A European survey of medical and paramedical professionals. *Progress in Brain Research, 177,* 329–338.

Di, H. B., Yu, S. M., Weng, X. C., Laureys, S., Yu, D., Li, J. Q., et al. (2007). Cerebral response to patient's own name in the vegetative and minimally conscious states. *Neurology, 68*(12), 895–899.

Esposito, F., Scarabino, T., Hyvarinen, A., Himberg, J., Formisano, E., Comani, S., et al. (2005). Independent component analysis of fMRI group studies by self-organizing clustering. *NeuroImage, 25,* 193–205.

Fernandez-Espejo, D., Junque, C., Vendrell, P., Bernabeu, M., Roig, T., Bargallo, N., et al. (2008). Cerebral response to speech in vegetative and minimally conscious states after traumatic brain injury. *Brain Injury, 22*(11), 882–890.

Giacino, J. T. (2004a). The vegetative and minimally conscious states: Consensus-based criteria for establishing diagnosis and prognosis. *NeuroRehabilitation, 19*(4), 293–298.

Giacino, J. T., Kalmar, K., & Whyte, J. (2004b). The JFK Coma Recovery Scale – Revised: Measurement characteristics and diagnostic utility. *Archives of Physical and Medical Rehabilitation, 85*(12), 2020–2029.

Giacino, J. T., Ashwal, S., Childs, N., Cranford, R., Jennett, B., Katz, D. I., et al. (2002). The minimally conscious state: Definition and diagnostic criteria. *Neurology, 58*(3), 349–353.

Gray, M. A., Minati, L., Harrison, N. A., Gianaros, P. J., Napadow, V., & Critchley, H. D. (2009). Physiological recordings: Basic concepts and implementation during functional magnetic resonance imaging. *NeuroImage, 47,* 1105–1115.

Greicius, M. D., Supekar, K., Menon, V., & Dougherty, R. F. (2009). Resting-state functional connectivity reflects structural connectivity in the default mode network. *Cerebral Cortex, 19,* 72–78.

Hauk, O., Johnsrude, I., & Pulvermuller, F. (2004). Somatotopic representation of action words in human motor and premotor cortex. *Neuron, 41*(2), 301–307.

Kwong, K. K., Belliveau, J. W., Chesler, D. A., Goldberg, I. E., Weisskoff, R. M., Poncelet, B. P., et al. (1992). Dynamic magnetic resonance imaging of human brain activity during primary sensory stimulation. *Proceedings of the National Academy of Sciences of the United States of America, 89,* 5675–5679.

Laureys, S. (2005). The neural correlate of (un)awareness: Lessons from the vegetative state. *Trends in Cognitive Sciences, 9*(12), 556–559.

Laureys, S., Berré, J., & Goldman, S. (2001). Cerebral function in coma, vegetative state, minimally conscious state, locked-in syndrome and brain death. In J. L. Vincent (Ed.), 2001 Yearbook of intensive care and emergency medicine. (pp. 386–396). Berlin: Springer-Verlag.

Laureys, S., Faymonville, M. E., Degueldre, C., Fiore, G. D., Damas, P., Lambermont, B., et al. (2000a). Auditory processing in the vegetative state. *Brain, 123*(8), 1589–1601.

Laureys, S., Faymonville, M. E., Luxen, A., Lamy, M., Franck, G., & Maquet, P. (2000b). Restoration of tha-lamocortical connectivity after recovery from persistent vegetative state. *Lancet, 355*(9217), 1790–1791.

Laureys, S., Faymonville, M. E., Moonen, G., Luxen, A., & Maquet, P. (2000c). PET scanning and neuronal loss in acute vegetative state. *Lancet, 355,* 1825–1826.

Laureys, S., Goldman, S., Phillips, C., Van Bogaert, P., Aerts, J., Luxen, A., et al. (1999a). Impaired effective cortical connectivity in vegetative state. *NeuroImage, 9*(4), 377–382.

Laureys, S., Lemaire, C., Maquet, P., Phillips, C., & Franck, G. (1999b). Cerebral metabolism during vegetative state and after recovery to consciousness. *Journal of Neurology, Neurosurgery & Psychiatry, 67*(1), 121.

Laureys, S., Owen, A. M., & Schiff, N. D. (2004a). Brain function in coma, vegetative state, and related disorders. *Lancet Neurology, 3*(9), 537–546.

Laureys, S., Perrin, F., Faymonville, M. E., Schnakers, C., Boly, M., Bartsch, V., et al. (2004b). Cerebral processing in the minimally conscious state. *Neurology, 63*(5), 916–918.

Levy, D. E., Sidtis, J. J., Rottenberg, D. A., Jarden, J. O., Strother, S. C., Dhawan, V., et al. (1987). Differences in cerebral blood flow and glucose utilization in vegetative versus locked-in patients. *Annals of Neurology, 22*(6), 673–682.

Logothetis, N. K., Pauls, J., Augath, M., Trinath, T., & Oeltermann, A. (2001). Neurophysiological investigation of the basis of the fMRI signal. *Nature, 412,* 150–157.

Maquet, P., Dive, D., Salmon, E., Sadzot, B., Franco, G., Poirrier, R., et al. (1990). Cerebral glucose utilization during sleep–wake cycle in man determined by positron emission tomography and [18F]2-fluoro-2-deoxy-D-glucose method. *Brain Research, 513*(1), 136–143.

Menon, D. K., Owen, A. M., Williams, E. J., Minhas, P. S., Allen, C. M., Boniface, S. J., et al. (1998). Cortical processing in persistent vegetative state. *Lancet, 352*(9123), 200.

Moritz, C. H., Rowley, H. A., Haughton, V. M., Swartz, K. R., Jones, J., & Badie, B. (2001). Functional MR imaging assessment of a non-responsive brain injured patient. *Magnetic Resonance Imaging, 19*(8), 1129–1132.

Multi-Society Task Force on PVS, The (1994). Medical aspects of the persistent vegetative state (1). *New England Journal of Medicine, 330*(21), 1499–1508.

Nakayama, N., Okumura, A., Shinoda, J., Nakashima, T., & Iwama, T. (2006). Relationship between regional cerebral metabolism and consciousness disturbance in traumatic diffuse brain injury without large focal lesions: An FDG–PET study with statistical parametric mapping analysis. *Journal of Neurology, Neurosurgery & Psychiatry, 77*(7), 856–862.

Onofrj, M., Thomas, A., Paci, C., Scesi, M., & Tombari, R. (1997). Event related potentials recorded in patients with locked-in syndrome. *Journal of Neurology, Neurosurgery & Psychiatry, 63*(6), 759–764.

Owen, A., Coleman, M., Boly, M., Davis, M., Laureys, S., Jolles, J., et al. (2007). Response to comments on "Detecting awareness in the vegetative state". *Science, 315*(5816), 1221.

Owen, A. M., Coleman, M. R., Boly, M., Davis, M. H., Laureys, S., & Pickard, J. D. (2006). Detecting awareness in the vegetative state. *Science, 313*(5792), 1402.

Owen, A. M., Menon, D. K., Johnsrude, I. S., Bor, D., Scott, S. K., Manly, T., et al. (2002). Detecting residual cognitive function in persistent vegetative state. *Neurocase, 8*(5), 394–403.

Perrin, F., Schnakers, C., Schabus, M., Degueldre, C., Goldman, S., Bredart, S., et al. (2006). Brain response to one's own name in vegetative state, minimally conscious state, and locked-in syndrome. *Archives of Neurology*, *63*(4), 562–569.

Posner, J., Saper, C., Schiff, N., & Plum, F. (2007). *Plum and Posner's diagnosis of stupor and coma*. New York: Oxford University Press.

Raichle, M. E. (2006). Neuroscience: The brain's dark energy. *Science*, *314*(5803), 1249–1250.

Raichle, M. E., MacLeod, A. M., Snyder, A. Z., Powers, W. J., Gusnard, D. A., & Shulman, G. L. (2001). A default mode of brain function. *Proceedings of the National Academy of Sciences of the United States of America*, *98*(2), 676–682.

Rudolf, J., Ghaemi, M., Haupt, W. F., Szelics, B., & Heiss, W. D. (1999). Cerebral glucose metabolism in acute and persistent vegetative state. *Journal of Neurosurgical Anesthesiology*, *11*(1), 17–24.

Schiff, N. D., Rodriguez-Moreno, D., Kamal, A., Kim, K. H., Giacino, J. T., Plum, F., et al. (2005). fMRI reveals large-scale network activation in minimally conscious patients. *Neurology*, *64*(3), 514–523.

Schnakers, C., Majerus, S., Goldman, S., Boly, M., Van Eeckhout, P., Gay, S., et al. (2008a). Cognitive function in the locked-in syndrome. *Journal of Neurology*, *255*(3), 323–330.

Schnakers, C., Perrin, F., Schabus, M., Hustinx, R., Majerus, S., Moonen, G., et al. (2009a). Detecting consciousness in a total locked-in syndrome: An active event-related paradigm. *Neurocase*, *15*, 271–277.

Schnakers, C., Perrin, F., Schabus, M., Majerus, S., Ledoux, D., Damas, P., et al. (2008b). Voluntary brain processing in disorders of consciousness. *Neurology*, *71*, 1614–1620.

Schnakers, C., Vanhaudenhuyse, A., Giacino, J., Ventura, M., Boly, M., Majerus, S., et al. (2009b). Diagnostic accuracy of the vegetative and minimally conscious state: Clinical consensus versus standardized neurobehavioral assessment. *BMC Neurology*, *9*, 35.

Smart, C. M., Giacino, J. T., Cullen, T., Moreno, D. R., Hirsch, J., Schiff, N. D., et al. (2008). A case of locked-in syndrome complicated by central deafness. *Nature Clinical Practice Neurology*, *4*(8), 448–453.

Soddu, A., Boly, M., Nir, Y., Noirhomme, Q., Vanhaudenhuyse, A., Demertzi, A., et al. (2009). Reaching across the abyss: Recent advances in functional magnetic resonance imaging and their potential relevance to disorders of consciousness. *Progress in Brain Research*, *177*, 261–74.

Staffen, W., Kronbichler, M., Aichhorn, M., Mair, A., & Ladurner, G. (2006). Selective brain activity in response to one's own name in the persistent vegetative state. *Journal of Neurology, Neurosurgery & Psychiatry*, *77*(12), 1383–1384.

Tommasino, C., Grana, C., Lucignani, G., Torri, G., & Fazio, F. (1995). Regional cerebral metabolism of glucose in comatose and vegetative state patients. *Journal of Neurosurgical Anesthesiology*, *7*(2), 109–116.

van Buuren, M., Gladwin, T. E., Zandbelt, B. B., van den Heuvel, M., Ramsey, N. F., Kahn, R. S., et al. (2009). Cardiorespiratory effects on default-mode network activity as measured with fMRI. *Human Brain Mapping*, *30*(9), 3031–3042.

Vanhaudenhuyse, A., Laureys, S., & Perrin, F. (2008a). Cognitive event-related potentials in comatose and postcomatose states. *Neurocritical Care*, *8*(2), 262–270.

Vanhaudenhuyse, A., Schnakers, C., Bredart, S., & Laureys, S. (2008b). Assessment of visual pursuit in post-comatose states: Use a mirror. *Journal of Neurology, Neurosurgery & Psychiatry*, *79*(2), 223.

Ylipaavalniemi, J., & Vigario, R. (2008). Analyzing consistency of independent components: An fMRI illustration. *NeuroImage 39*, 169–180.

COGNITIVE NEUROSCIENCE, 2010, 1 (3), 204–240 (Includes Discussion paper, commentaries, and response.)

Discussion Paper

How neuroscience will change our view on consciousness

Victor A. F. Lamme

University of Amsterdam, Amsterdam, The Netherlands

Is there consciousness in machines? Or in animals? What happens to consciousness when we are asleep, or in vegetative state? These are just a few examples of the many questions about consciousness that are troubling scientists and laypersons alike. Moreover, these questions share a striking feature: They seem to have been around forever, yet neither science nor philosophy has been able to provide an answer. Why is that? In my view, the main reason is that the study of consciousness is dominated by what we know from introspection and behavior. This has fooled us into thinking that we know what we are conscious of. The scientific equivalent of this is Global Workspace theory. But in fact we don't know what we are conscious of, as I will explain from a simple experiment in visual perception. Once we acknowledge that, it is clear that we need other evidence about the presence or absence of a conscious sensation than introspection or behavior. Assuming the brain has something to do with it, I will demonstrate how arguments from neuroscience, together with theoretical and ontological arguments, can help us resolve what the exact nature of our conscious sensation is. It turns out that we see much more than we think, and that Global Workspace theory is all about access but not about seeing. The exercise is an example of how neuroscience will move us away from psychological intuitions about consciousness, and hence depict a notion of consciousness that may go against our deepest conviction: "My consciousness is mine, and mine alone." It's not.

Keywords: Neuroscience; Consciousness; Qualia; Global workspace; Visual perception; Re-entrant.

A STALEMATE BETWEEN INTROSPECTION AND BEHAVIOR

This paper is about seeing. So look at the scene of Figure 1, just briefly. What did you see? You are probably aware that there were colored objects, arranged in a circle. You may remember some colors, red and green, some shapes or even the identity of some objects, like the bread or the motorcycle. While you are reminiscing, you will undoubtedly also get the impression that you are forgetting. Somehow, the visual experience seems to fade away, from the rich and detailed representation you had at the moment you looked, to the impoverished, almost verbal trace that you are pondering over now. So what were you really seeing, what was in your conscious mind *at the moment* you looked? When you look again, the idea will no doubt settle in your mind that during the looking itself your visual sensation is in fact pretty rich: you see the whole picture. Sort of.

What is evident from this small exercise is that introspection is a poor guide to conscious visual sensation. To "know" what you are seeing, you need

Correspondence should be addressed to: Victor A. F. Lamme, Department of Psychology, University of Amsterdam, Roetersstraat 15, 1018WB Amsterdam, The Netherlands. E-mail: v.a.f.lamme@uva.nl

Victor Lamme is supported by an Advanced Investigator grant from the European Research Council, and by a grant from the Netherlands Foundations for Scientific Research (NWO).

DOI: 10.1080/17588921003731586

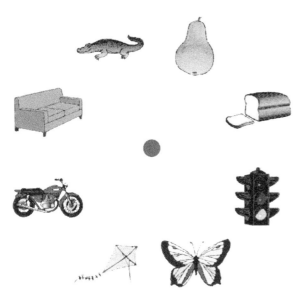

Figure 1. Look at this image for a second, and try to figure out what you saw.

to resort to cognitive functions such as attention, memory, and verbalization. At the same time, you notice that these functions impose a limit on the transfer from visual sensation to cognition. They do not seem to capture your full conscious experience. So which is it? Do I see what I see, or do I see what I know I am seeing?

Psychology to the rescue. Can't we design a task in order to really know how much we are seeing? A classic way to do such a thing would be the *whole report* paradigm. Subjects see an array of items, and the task is to name as many as possible immediately after the array has disappeared (Sperling, 1960). What is found in such experiments is that the capacity to store and retrieve even simple items, such as letters or numbers, is disappointingly low. The average subject will typically only reliably report about four objects. A somewhat more modern way of gauging the capacity of conscious vision is the *change detection* paradigm (Simons & Rensink, 2005). Here an array of items is shown as well, but the task of the subject is to compare that array to a second array that is presented after the first one, with a brief interruption in between (see Figure 2a). Typically, only one item of the array has changed between the first and second display in 50% of the trials, while in the other 50% there was no change at all. The task of the subject is to indicate whether a change was present or not, or even more simply, to indicate whether the object that is cued in the second array has changed or not. Surprisingly, subjects perform very poorly on this task (Landman, Spekreijse, & Lamme, 2003a, 2004b; Sligte, Scholte,

& Lamme, 2008), typically only moderately above chance level (~70%). Percentage correct, in combination with the number of items that has been used, can be converted to the number of items that subjects can simultaneously monitor for change; in other words, the capacity of consciously accessible information that is extracted from the scene. Change detection paradigms come in many flavors (Simons & Rensink, 2005), but in all cases the capacity of conscious access that comes out of these experiments is rarely higher than four objects (Figure 2f; Sligte et al., 2008), and it may even drop to (much) lower values depending on the exact paradigm, or the complexity of the objects (Figure 2c).

The picture of conscious vision that emerges from these findings is that of limitation. Apparently, consciousness is fairly sparse, and much more limited than our introspection would suggest. We may think we see a full and detailed image of what is in front of our eyes, but we take in only a small subset of the information (Dehaene & Naccache, 2001; O'Regan & Noe, 2001). This is all the more evident when we look at other paradigms that have been used to gauge the capacity of conscious access, such as the *inattentional blindness* experiments, where the appearance of large and fully visible objects is missed when attention is focused on something else (Mack & Rock, 1998), or *attentional blink* experiments where items are missed when another item is detected briefly before (Sergent, Baillet, & Dehaene, 2005; Shapiro, 2009; Shapiro, Raymond, & Arnell, 1994). Introspectively, consciousness seems rich in content. We see colors, shapes, objects, and seemingly everything that is in front of us. From the third-person perspective of the behavioral scientist, however, consciousness is rather miserable. Which is the truth (Block, 2005)?

A RICHER REPRESENTATION?

One might argue that these paradigms measure not so much the capacity of conscious vision, but rather the capacity of working memory, attention or any other function necessary for access. It could be that the limit sits in one of these functions, not so much in the visual sensation itself. They are necessary to complete the behavioral task, and if they have limited capacity they will impose a limit on a potentially much richer visual sensation. There is indeed evidence that this might be the case. Many decades ago, it was shown that the whole report paradigm can be modified slightly to yield a completely different result. Instead of asking subjects what they saw, Sperling (1960) used a cue to point at the location of items that had to

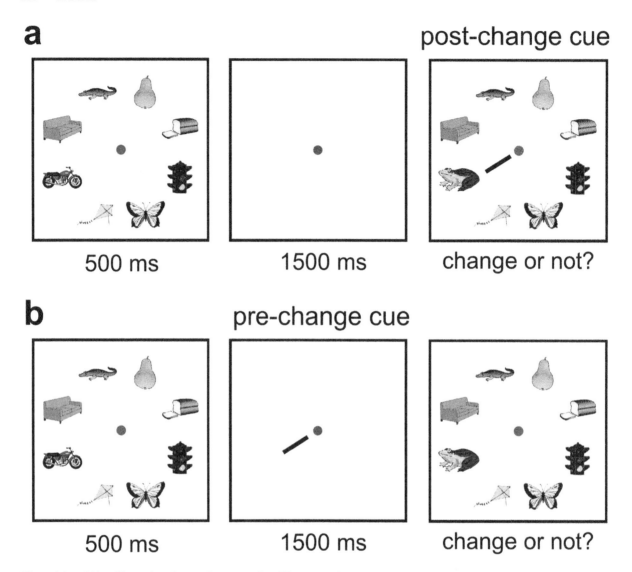

Figure 2 (a and b). Change detection paradigms, revealing different capacity representations. (a) The traditional change blindness paradigm, where an array of objects is presented for 0.5 s or so, followed by a blank screen for several seconds, followed by a second array. The task of the subject is to indicate whether the cued object has changed or not (in this case it changes, but in 50% of trials it doesn't). Subjects perform poorly in this task, and the percentage correct can be converted to a capacity, which in this case is about two objects that can be monitored for change (see result in panel c). (b) The same paradigm, but now with a cue somewhere during the interval. This yields a much higher percentage correct, and hence a higher capacity (see results in panel c).

be reported. He presented three rows of four letters, and after disappearance of the array a tone indicated to subjects whether they had to report the letters of the top, middle, or bottom row. Again, only three or so letters were reported. But it didn't matter what row was cued. That implies that at the moment of the cue (up to a second after presentation) the subject must have had a representation in his mind that was much richer than the three items that were reported—in fact three times (the number of rows) as large, i.e. about nine letters. This richer representation was named *iconic memory*, and many subsequent *partial report*

experiments revealed that it consisted of an almost-carbon copy of what is presented to the subject (Coltheart, 1980).

We (and others; Becker, Pashler, & Anstis, 2000)) combined the change blindness and partial report paradigms in a single experiment to compare the different representations directly (Figures 2a and 2b). An array of objects is presented, followed by a blank interval of up to several seconds, again followed by the array. One of the objects might have changed, and a cue points to that object. Largely different results are obtained depending on when exactly the cue is presented

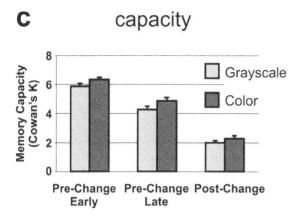

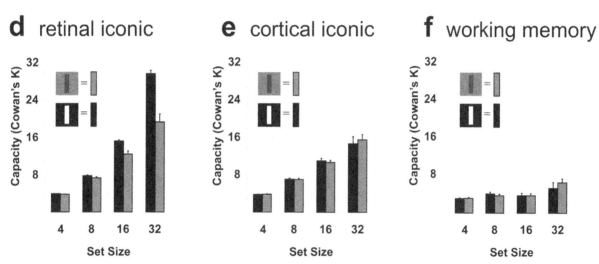

Figure 2 (c, d, e and f). (c) Results for the change blindness experiments shown in panels a and b. Capacity is high for the pre-change cues (i.e., paradigm b), and low for the post-change cues (paradigm a). Using b/w or colored objects makes no difference. (d) Results for change detection using a pre-change cue immediately after an array of randomly oriented rectangles has disappeared. Capacity increases with set size, i.e., the number of objects used. Black bars are for white rectangles on a black background, gray bars for isoluminant red on gray. These results probably reflect a retinal afterimage. (e) Results from the same paradigm as in panel d, but with a cue 1 s after disappearance of the first array. Results probably reflect a cortical "after-image". (f) Results from the same paradigm as in panel d, but with a post-change cue. Results reflect the (fixed) capacity of working memory.

(Figure 2c). Presenting the cue together with the first array makes the task fairly trivial, as this directs attention to the items that might change—a change that is not difficult to detect in itself. A cue together with the second array yields the low performance and limited capacity of change detection results. However, a cue presented during the interval, even as long as 1 or 2 s after disappearance of the first array, yields a capacity that is much larger, almost as large as when the cue is shown with the first array still there (Landman et al., 2003a, 2004b; Sligte et al., 2008). This confirms and extends the partial report/iconic memory results. Large capacity representations are found even seconds after disappearance of the first array, provided the second array is not allowed to overwrite these.

Research using this combined paradigm so far has shown the large capacity/iconic memory representa-

tion to be of fairly long duration (up to 4 s), and to be quasi-linearly dependent on the number of items that are presented; the more items are shown, the more are stored, but probably with a plateau (Sligte et al., 2008; see Figure 2e). Capacity also depends on the complexity of objects. Presenting a new scene erases (or interferes with) this form of iconic memory, and this has been confirmed using other intervals than the blank screen shown in Figure 2. Critical for this interference is the presentation of objects at the same location. For example, a homogeneous texture of oriented line segments has no effect, but when these line segments compose figures at the same location as the objects of the memory array, performance drops to the low capacity of working memory. However, objects presented at other locations pose few problems. Even when an object is shown at fixation during the interval there is little interference, suggesting that the large capacity is not erased by the capturing of attention (Landman et al., 2004b). This is confirmed by an experiment where the cue during the interval was followed in some trials by a second cue later in that interval. The capacity of the representation addressed by this second cue is equal to the capacity addressed by the first cue, i.e., the focusing of attention on the item in iconic memory does not make the iconic memory representation 'collapse' (Landman et al., 2003a).

Some controversy exists as to whether the large capacity representation that is measured using the cued change detection task is identical to iconic memory as it was classically defined, or maybe is better seen as some fragile form of working memory. Indeed different forms exist. Immediately after presentation of the array, a cue will reveal an almost exact carbon-copy of what is on the retina, which is erased by any light or colored screen (Sligte et al., 2008). This is probably a retinal after-image (Figure 2d). However, the large capacity store that is obtained with a cue several seconds after disappearance (which is what is discussed above) is not that easily erased, and there are sufficient arguments against it being a mere retinal after-image. It is probably best viewed as a "cortical after-image", i.e., the neural activity that remains in the visual brain after a visual stimulus has been removed. A recent neuroimaging study showed that the neural correlate of an item being part of the large capacity (iconic) representation is neural activity in visual area V4 (Sligte et al., 2009).

WHAT ARE WE SEEING?

Whether this type of iconic memory is better viewed as a fragile form of working memory is not relevant for the discussion here. The key question is this: These results force us to acknowledge that different neural representations of a scene exist (or at least, different representations remain immediately after that scene has disappeared). There is a stable representation, linked to working memory and attention, that allows access and report, and can be maintained across views, yet has fairly limited capacity. And there is a much larger capacity representation—perhaps a virtual copy of the outside world—that is however very fragile, fading away in a few seconds and overwritten as new visual information hits the eyes (Figure 3a). Which is the conscious one? Which is better evidence of what we are—or rather were[1]—seeing when we look at a scene (Block, 2007)?

Face value arguments don't really resolve the issue. Introspectively, it is almost impossible to know

whether you really see all the objects of Figure 1, or just the few you focus attention on. As has been argued, it could be that the impression of seeing all objects is an illusion, created by the fact that every time you focus on one of the objects it is there (often referred to as the refrigerator light illusion: When you open the door it's always on, but in reality it is not; Dehaene et al., 2006). Behaviorally, we can only argue for the presence of two representations, a limited and a large capacity one. Whether there is phenomenality (i.e., a conscious sensation) in either of the two is not directly addressed by the experiment.

Deciding whether there is phenomenality in a mental representation implies putting a boundary—drawing a line—between different types of representations (regardless of whether that is a sharp or a fuzzy boundary). Where to draw that line cannot be decided on the basis of this experiment alone. We have to start from the intuition that consciousness (in the phenomenal sense) exists, and is a mental function in its own right. That intuition immediately implies that there is also *unconscious* information processing. These intuitions come from the extreme ends: cases where the presence or absence of conscious sensations is undisputed. For example, when I show a picture of a face to someone and he replies by confirming to see it, verbally sketching an accurate outline, giving a description of the texture of the face, the emotional expression, and every other feature of it, there is little reason to doubt the presence of a conscious sensation of that face. Likewise, when I show that picture for only 5 ms, followed by a strong mask, and then the subject is incapable of telling whether an image was presented at all, incapable of making a higher than chance forced choice guess whether it was a face or a house, black or white, or bearing any other feature, let alone giving a description of the identity or expression of that face, there is little reason to doubt the absence of a conscious sensation. Especially when the subject is fully focusing his attention on the location where the image is shown, has no neurological disorders, or any other condition that would logically prevent a conscious sensation—if present—to be reported, the only logical conclusion would be to infer the absence of such a sensation.

From that starting point, we infer phenomenality in other situations, or not. A conservative approach would be to infer phenomenality only when the subject confirms having it. This runs into the problem of what

---

[1] Please note that neither iconic nor working memory representations have any visual phenomenality themselves. They are just memory traces, of course. But they are the closest evidence of what you were seeing when the image was still there, and as such give us a window on the representations that are present *during* the seeing.

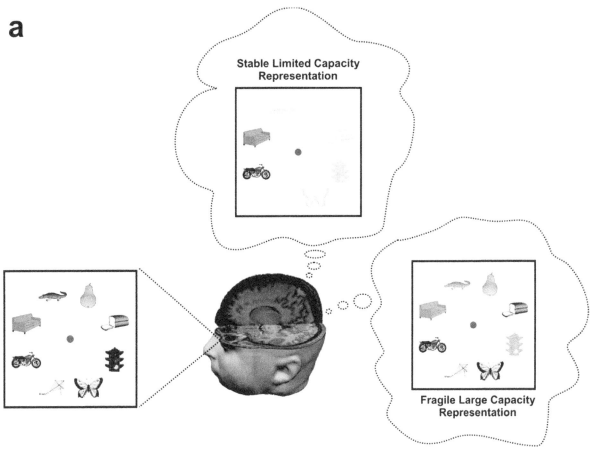

a

Stable Limited Capacity
Representation

Fragile Large Capacity
Representation

Which representation is the conscious one?

b

	Iconic Memory	**Working Memory**
Capacity	Large: scales with number of objects in scene, up to 16 documented	Limited: typically 4 or less, depending on complexity
Duration	Short: Up to 4 seconds	Medium: Minutes to hours
Stability	Fragile: Overwritten by any new scene containing objects at approx. the same location. Probably also erased by eye movements.	Stable: Resistant to new visual information, eye movements. Interference from other working memory load
Frame of reference	Mostly retinotopic	Mostly spatiotopic
Quality	Some feature binding, high visual accuracy. Capacity drops with complexity of objects	Feature binding, 'object files', etc. capacity drops with complexity of objects
Cognitive value	Limited	Extensive

Figure 3.   (a) Two representations are generated by our brains when we see a scene, one fragile with large capacity, one stable with limited capacity. (b) Properties of the two representations.

should be considered proper "confirmation." If every behavioral measurement—verbal report, yes–no detection, forced choice discrimination, pointing, priming, etc.—indicates its presence, there is little problem. But what if some behaviors indicate "present" while others do not? This happens all the time when we study consciousness, either by manipulating it (masking, change blindness, attentional blink, rivalry, etc.) or when it has been altered by neurological conditions (blindsight, split-brain, neglect, extinction, etc.). Then some—essentially—arbitrary choice has to be made as to what type of behavior constitutes proper evidence for the presence of conscious sensation and what does not (Lamme, 2006; Seth, Dienes, Cleeremans, Overgaard, & Pessoa, 2008). An alternative would be to take a more liberal approach, and to assign phenomenality whenever there is no proof of absence, i.e., when any behavioral measurement gives a green light. In that case, however, one immediately runs into the problem of drawing a line with the unconscious, as *some* response, such as an altered skin conductance or pupil dilatation, is always present (Stoerig, 1996).

In the situation we focus on here, the two representations discussed above, we face a similar problem. There is little reason to doubt that there is (was) phenomenality in the limited capacity working memory representation. The issue hinges on whether there is any such thing in the iconic memory representation. Should we group that with the conscious, or rather with the unconscious?

FUNCTIONAL ARGUMENTS

A way to answer this question may come from functional arguments. Soon after the discovery of the high capacity representation of iconic memory, it was discarded as having no functional use (Haber, 1983). Seemingly, any cognitive manipulation of visual input requires it to be accessed by working memory or attention. And that imposes the limits discussed above. Also the instability of iconic memory—it being overwritten as soon as we move our eyes to a new fixation—argues against it being of any cognitive use (Figure 3b). So if the *raison d'être* of consciousness is cognitive access to the information, and the ability to cognitively manipulate that information, and combine it with information stored in working or long-term memory, or with inputs from other senses, then the most sensible solution would be not to view the iconic representation as part of consciousness. A theoretical framework that supports this idea is the *Global Workspace* model of consciousness (Baars, 2005).

There are, however, two arguments against this. First, by linking consciousness so much with cognition, there is some "throwing away of the baby with the bathwater," because cognition and access do very little to explain the key feature of consciousness that we consider here, which is phenomenality. Why would combining visual input with working memory make it "visible" to the mind's eye—in other words, produce qualia? That makes me smell a homunculus, in the sense that visual information seemingly needs to *go somewhere* to achieve phenomenality. Second, recent experiments force us to acknowledge that also in the unconscious there is a lot of cognition going on, such as multisensory integration (de Gelder, Pourtois, & Weiskrantz, 2002), interaction with long- and short-term memory (Schacter, Chiu, & Ochsner, 1993; Watanabe, Nanez, & Sasaki, 2001), cognitive control (Lau & Passingham, 2007; van Gaal, Ridderinkhof, Fahrenfort, Scholte, & Lamme, 2008; van Gaal, Ridderinkhof, van den Wildenberg, & Lamme, 2009), attention and selection (Kentridge, Heywood, & Weiskrantz, 1999; Koch & Tsuchiya, 2007), and even reasoning and thinking (Bechara, Damasio, Tranel, & Damasio, 1997; Dijksterhuis, Bos, Nordgren, & van Baaren, 2006). That makes these functional arguments intrinsically ill-suited to put a sharp divide between conscious and unconscious processing.[2]

PHENOMENAL QUALITIES

A better approach to answer the question would be to study the phenomenal qualities of the two representations. Much is known about such qualities in the case of accessible and reportable percepts. For example, conscious percepts typically show the integration of features; in other words, grouping and binding. In a conscious percept, the loose set of elements that make up a scene are typically grouped into coherent surfaces and objects (Nakayama & Shimojo, 1992; Serences & Yantis, 2006; Singer, 1999). Moreover, different features, such as color, shape, motion, and size, are linked to each other. For example, when consciously seeing (and attending) the motorcycle of Figure 1, you instantaneously also see that it is red, and is

---

[2] Functions, whether cognitive or not, are of course also seen as irrelevant to consciousness in the original formulation of the so called *hard problem* of consciousness (Chalmers, 1995). I am not implying that that line of reasoning should be followed fully, as that way of posing the problem makes phenomenality—or qualia—almost impossible to study. For example, it renders invalid the very intuitions on which the conscious–unconscious divide is based (see above). But I do agree that many functions—cognitive functions in particular—do very little towards explaining qualia.

TABLE 1
Which key aspects of phenomenality (feature binding, inference, competition) are present in unattended or unreported visual representations?

	Iconic memory/Change blindness	Inattentional blindness	Attentional blink	Neglect extinction	Split brain, nondominant hemisphere
Feature binding or fig–ground segregation	Landman et al., 2003a, 2003b, 2004b	Scholte et al., 2006			
Object recognition		Mack & Rock, 1998	Marois et al., 2004		Sperry, 1984
Inference (illusions)				Vuilleumier et al., 2001	Corballis, 2003
Competition (rivalry)		Lee et al., 2007			O'Shea & Corballis, 2003

Notes: The table lists reported evidence for the presence of these features in various conditions that are characterized by the absence of attention and report. References are not meant to be exhaustive. Blank fields indicate the absence of knowledge, i.e., a fruitful area of further research. Object recognition is added to the list for completeness, but it may be questioned whether this is a feature of phenomenality, as categorical discrimination is also present in clearly unconscious states, such as masking or blindsight.

not wearing the blue of the couch next to it. We know that grouping and binding exist in working memory, and there is a strong tendency to believe they depend on attention (Luck & Vogel, 1997; Treisman, 1996). However, it has been shown that also in iconic memory there is figure–ground segregation of oriented line segments (Landman, Spekreijse, & Lamme, 2004a; Landman et al., 2004b), and binding of features such as size and orientation (Landman et al., 2003a).

Another property of conscious percepts is that they often are the end result of a competition between several possible groupings and bindings. This is most prominent in bi-stable phenomena such as perceptual or binocular rivalry (Blake & Logothetis, 2002). In addition, perception often is nonveridical, in that inferences are drawn that move away from the physical input. This is evident from the many visual illusions that exist, where we are tricked into seeing things that are not "really" there (e.g., Churchland & Churchland, 2002). It is unclear whether these phenomena—perceptual competition and inference—also exist when we do not attend to the visual input, such as in the case of the iconic memory representation, or during inattentional blindness or change blindness. It is known that both phenomena are largely independent of attention: Rivalry switches cannot be fully suppressed at will (Meng & Tong, 2004), and visual illusions are the prime example of the cognitive impenetrability of visual perception (Pylyshyn, 1999). Even when you know that the two double arrows of the Muller-Lyer illusion are of equal length, you still see them as different (Bruno & Franz, 2009). It is interesting to note that in visual neglect or extinction, visual illusions coming from the neglected hemifield still "work" to influence the percepts that are reported by the patient from the intact hemifield (Vuilleumier,

Valenza, & Landis, 2001). That is a first example that phenomenal qualities exist in a representation that is inaccessible by the patient.

Studying the phenomenal qualities of the two different representations would be a good research agenda to answer the question of whether there is phenomenality in iconic memory. If it turns out that the iconic memory representation shares almost all phenomenal qualities with attended/working memory representations—except cognitive access—it would make the most scientific sense to acknowledge phenomenality to this large capacity representation (and conversely to conclude that our phenomenal experience is widespread rather than limited). However, at present this is unclear. Similarly, it would be useful to study precisely the phenomenal qualities of other states where attention or report is absent, such as inattentional blindness, change blindness, attentional blink, neglect, or split-brain. In some cases, evidence for phenomenal qualities in those conditions has already been reported (see Table 1). What if we could fill the whole table? Shouldn't we conclude that the existence of phenomenality without report is the more parsimonious conclusion?

NEURAL ARGUMENTS

Would it help to find the neural correlates of the two representations (Crick & Koch, 1998a; Crick & Koch, 2003; Rees, Kreiman, & Koch, 2002)? Seemingly not. If we know that the large capacity iconic memory representations sits in visual cortex, while the limited working memory representation depends on the fronto-parietal network, how could that ever answer whether there is phenomenality to either of them? The explanatory gap only seems larger between neuroscience and phenomenality. Neuroscience can, however, help to solve the seemingly impossible, and it can do so in different ways (Haynes, 2009; Rees, 2007; Tononi & Koch, 2008). My approach is to go beyond neural correlates, and turn these into neural arguments (Lamme, 2006; Seth, 2008; Seth et al., 2008; Seth, Izhikevich, Reeke, & Edelman, 2006).

The approach would be more or less equivalent to what is described above for the study of phenomenal qualities of the different representations. What if we could show that the neural correlate of the iconic memory representation shares all its essential qualities with the working memory representation—except the neural qualities that enable access and report? This would require finding the neural correlates of both types of representations—which is far less difficult than finding *the* neural correlate of consciousness—and finding out which are the essential qualities of these correlates that matter for phenomenality, which is almost as difficult as finding the neural correlate of consciousness but is a much more theoretical exercise (Seth et al., 2006; Tononi, 2004, 2008).

What would be the neural correlates of the two representations gauged via the paradigm of Figure 1? It has been shown (and explained at greater length in Lamme, 2004; Lamme & Roelfsema, 2000; Lamme, Super, & Spekreijse, 1998a) that each time we lay our eyes on a scene (by making an eye movement, or when it is flashed in an experiment), cortical visual processing goes through a succession of stages. First, information flows from visual to motor areas in what is called the fast feedforward sweep (FFS; Lamme & Roelfsema, 2000). Within 100 to 120 ms (in monkeys; probably about 200 ms in humans), activity spreads from V1 to the extrastriate and dorsal and ventral stream areas, all the way up to motor cortex, and prefrontal regions involved in controlling and executing movement. In some respects we could call this a cortical reflex arch, mediated by the feedforward connections. During the FFS, early visual areas extract features of the image such as orientation, shape, color, or motion (Bullier, 2001; Lamme & Roelfsema, 2000). But high level features are also detected. Cells in inferotemporal cortex distinguish between face and non-face stimuli with their first spikes (Oram & Perrett, 1992; Rolls & Tovee, 1994). The FFS thus enables a very rapid categorization of visual stimuli into all sorts of (probably) behaviorally relevant categories. Potentially related motor responses are initiated when the FFS reaches motor regions (Dehaene et al., 1998), and control centers are activated in prefrontal cortex (Lau & Passingham, 2007; van Gaal et al., 2008, 2009).

Not all stimuli travel all the way up. If multiple stimuli are presented, many can be represented at the early stages. However, in successively higher areas, competition between multiple stimuli arises. Attentional selection (in one way or another; Egeth & Yantis, 1997) may resolve this competition (Desimone, 1998; Desimone & Duncan, 1995). In the end, only a few stimuli reach the highest levels, such as the areas involved in planning and executing behavior. And because only attended stimuli are selected for deep processing, only these influence behavior, can be reported, are stored in working memory, etc. Unattended ones "die out" in the early stages of processing (Sergent et al., 2005). From a neural perspective, attention (or the consequence of attention) can thus be straightforwardly defined as the depth of processing that a stimulus reaches (Dehaene, Changeux, Naccache, Sackur, & Sergent, 2006; Lamme, 2003, 2004).

As soon as the FFS has reached a particular area, horizontal connections start to connect distant cells within that area, and feedback connections start sending information from higher level areas back to lower levels, even all the way down to V1 (Bullier, 2001; Salin & Bullier, 1995). Together, these connections provide what is called recurrent processing (RP) (Edelman, 1992; Lamme & Roelfsema, 2000), which is fundamentally different from the FFS. RP allows for dynamic interactions between areas that can grow ever more widespread as time after stimulus onset evolves. RP may thus form the basis of dynamic processes such as perceptual organization, where different aspects of objects and scenes are integrated into a coherent percept (Lamme & Spekreijse, 2000; Lamme et al., 1998a; Lamme, Vandijk, & Spekreijse, 1993; Sporns, Tononi, & Edelman, 1991). Also in RP, attentional selection makes a difference. Attentional amplification may turn RP into the widespread coactivation of visual and fronto-parietal areas, including parts of the brain that mediate action or control, so as to produce a coordinated and planned response to selected visual information (Dehaene et al., 2006).

The FFS is more or less automatically followed by RP. It is in fact very difficult to prevent FFS activation from being followed by RP. The only way to do this is by forcing a second FFS to follow the first one before RP related to the first one can fully ignite. This is what happens when a visual stimulus is masked: The FFS activation of the mask prevents RP for the target stimulus, which then becomes invisible (Enns & Di Lollo, 2000; Fahrenfort, Scholte, & Lamme, 2007; Lamme, Zipser, & Spekreijse, 2002).

FOUR STAGES OF NEURAL PROCESSING

Because depth of processing (attention) and the FFS/RP distinction are orthogonal, a visual stimulus can reach any of four stages of processing, illustrated in Figure 4.

- *Stage 1:* Superficial processing during the FFS. This would happen when a stimulus is not attended and masked. Unattended and masked words, for example, do not activate word-form selective areas, only visual areas (Dehaene et al., 2006), so do not even penetrate deeply into the ventral stream hierarchy.
- *Stage 2:* Deep processing during the FFS; for example, a stimulus that is attended, yet masked (and hence invisible). This stimulus does travel through the whole hierarchy of sensory to motor and prefrontal areas, and may influence behavior, as in unconscious priming (Dehaene et al., 1998; Eimer & Schlaghecken, 2003; Thompson & Schall, 1999).
- *Stage 3:* Superficial processing of a recurrent/re-entrant nature (RP); for example, a visual stimulus that is given sufficient time to evoke RP (i.e., is not masked within ~50 ms) yet is not attended or is neglected, as in neglect (Driver & Mattingley, 1998), inattentional blindness (Scholte, Witteveen, Spekreijse, & Lamme, 2006), change blindness (Landman, Spekreijse, & Lamme, 2003b; Landman et al., 2004a; Schankin & Wascher, 2007), or the attentional blink (Marois, Yi, & Chun, 2004).
- *Stage 4:* Deep (or a better word may be "widespread") RP. This is the case when RP spans the whole hierarchy from low level sensory to high level executive areas. This occurs when a stimulus is given sufficient time to engage in RP and is attended. Others have equated this to the situation that a stimulus has entered global workspace (Baars, 2005; Dehaene & Naccache, 2001).

How would this apply to glancing at the image of Figure 1? As soon as we look at the image, the FFS will be activated. Depending on where your attention happens to be focused, the activation from some objects will travel all the way up to motor and prefrontal areas, while others activate only some visual areas. Given that the image is not masked, this will then be followed by RP for all objects in the image. Attentional selection determines the extent of RP for each object. Objects that were attended during the FFS will have an advantage, but attentional selection can also switch to items that proved to be more salient during the FFS (i.e., penetrated more deeply, despite attention being focused elsewhere). Subsequently, many objects will evoke RP that is limited to a few visual areas (Stage 3) (Landman et al., 2003b), while only some evoke widespread RP (Stage 4), like in global workspace activation (Dehaene et al., 2006). Finally, the stimulus is removed. What then remains is the traces of the pattern of RP that was present at the moment the stimulus was switched off. Stage 3 turns into iconic memory, while Stage 4 turns into working memory (Figure 5).

The question we try to answer in this paper can now be propped up by neural arguments. We know that the large capacity representation of iconic memory corresponds to the remains of Stage 3 processing, while the limited capacity working memory representation corresponds to what remains of Stage 4. What would be the neural argument to grant phenomenality to Stage 3? Assuming there is phenomenality in Stage 4, and not in Stage 1 (remember the starting intuitions), we have to decide which are the essential qualities of Stage 4 that would "produce" the phenomenality, and see whether these neural qualities are also present in Stage 3. In Stage 4, we have recurrent processing and activation of fronto-parietal areas, which we don't have in Stage 1. Either of the two—or their combination—therefore is a candidate neural ingredient for phenomenality.

Let us first turn to the involvement of the frontoparietal network, a key ingredient of the neural equivalent of Global Workspace theory (Dehaene & Naccache, 2001). It is argued that activation of prefrontal areas is a prerequisite for consciousness (Rees, 2007) because it has the long range connections that enable the integration of information from otherwise widely separated regions of the brain (Del Cul, Dehaene, Reyes, Bravo, & Slachevsky, 2009). However, a convergence of information towards prefrontal cortex in itself does not appear to be sufficient for conscious sensations to arise. For example, there are now several studies showing activation of regions such as the frontal eye fields, anterior cingulate, pre-supplementary motor area, inferior frontal gyrus, and anterior insula by masked stimuli or other unconscious events (Klein et al., 2007; Lau & Passingham, 2007; Thompson & Schall, 1999; van Gaal et al., 2008, 2009). The type of masking used in these studies leaves little doubt about the absence of any visual sensation, as even forced choice detection and other rigorous behavioral measurements are at chance. Yet despite being unconscious, the prefrontal activation is functional, in the sense that it evokes effects of control on subsequent visible stimuli, such as response inhibition or strategic switching. Apparently, a feedforward convergence of information towards the

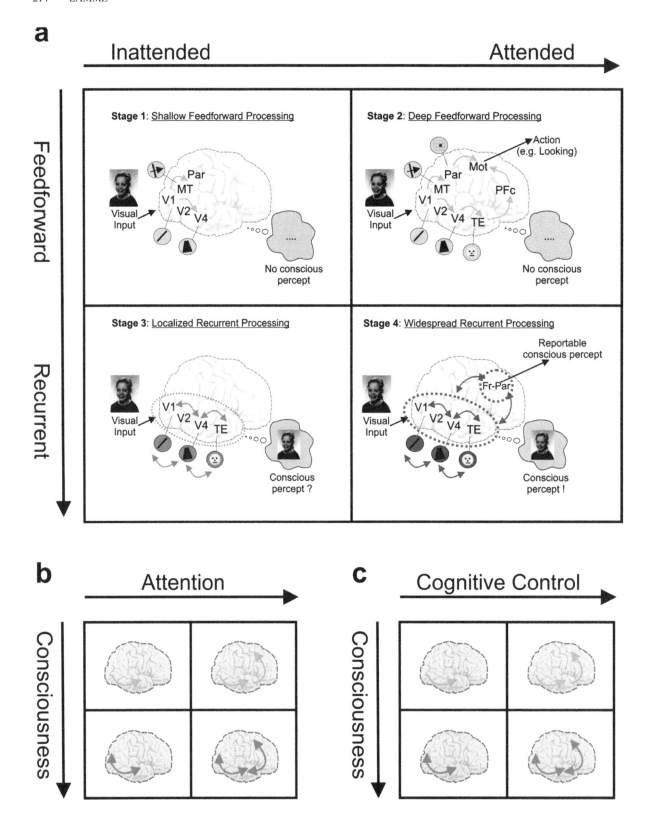

Figure 4.   (a) Four stages of cortical processing (see text for explanation). (b, c) When the difference between feedforward and recurrent processing (vertical axis, panel a) is identified to the difference between unconscious and conscious processing (vertical axis, panels b and c), how consciousness is orthogonal and independent of attention (b) and cognitive control (c) is readily explained.

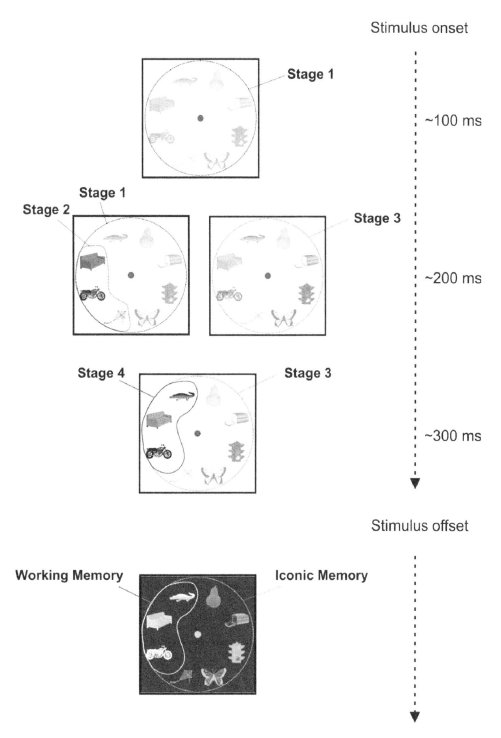

Figure 5. (a) A graphical depiction of the temporal evolution of processing stages that each of the objects in Figure 1 may reach at successive moments in time after presentation of the stimulus array, and how these processing stages change into iconic and working memory representations once the stimulus is removed. Initially, all objects are processed by low level areas in a feedforward fashion, so that basic features are extracted (Stage 1: faint gray shades). Some objects are processed more deeply (Stage 2: higher contrast gray shades), depending on top down and bottom up attentional selection. Meanwhile, recurrent processing in early visual areas emerges (Stage 3; faint colors) for all or most of the objects. Later still, recurrent processing grows more widespread (Stage 4, vivid colors) for those objects that are selected by attention (potentially slightly different ones than those that were favored initially, as attentional selection is influenced by previous processing). After stimulus removal, Stage 3 processing turns into iconic memory, while Stage 4 processing turns into working memory (inverted colors). In change detection paradigms (Figure 2), the sequence is repeated once the second array appears, and all stages are reset or overwritten, except for the Stage 4/working memory representation.

prefrontal cortex in itself is not yielding any conscious sensation, even when this information activates processes such as control and inhibition. In other words, both Stage 1 and Stage 2 processing are unconscious, no matter what areas are reached.

The remaining difference between Stage 4 and Stages 1 and 2 is that in the latter there is only feedforward processing, while in Stage 4 (and Stage 3) there is recurrent processing. Could that be the essential ingredient that gives phenomenality to Stage 4? And if so, what is so special about recurrent processing compared to feedforward processing that it yields conscious sensation? That recurrent processing is necessary for visual awareness is now fairly well established, and supported by numerous experiments (Boehler, Schoenfeld, Heinze, & Hopf, 2008; Camprodon, Zohary, Brodbeck, & Pascual-Leone, in press; Haynes, Driver, & Rees, 2005; Lamme, Super, Landman, Roelfsema, & Spekreijse, 2000; Lamme, Zipser, & Spekreijse, 1998b; Lamme et al., 2002; Pascual-Leone & Walsh, 2001; Silvanto, Cowey, Lavie, & Walsh, 2005; Super, Spekreijse, & Lamme, 2001). The key issue here is to what extent recurrent processing can be considered the ingredient of Stage 4 that *explains* phenomenality, and whether it does so better than the other main ingredient, which is the involvement of the fronto-parietal network.

KEY PROPERTIES OF PERCEPTION EXPLAINED BY RECURRENT PROCESSING

It is well established that many computational problems in vision can be solved better using recurrent rather than feedforward processing architectures. Among the many examples are feature grouping and segregation, figure–ground assignment, depth sorting of surfaces, and occlusion (Grossberg & Pessoa, 1998; Jehee, Lamme, & Roelfsema, 2007a; Roelfsema, Lamme, Spekreijse, & Bosch, 2002; Thielscher & Neumann, 2008). These processes are the first steps in going from the features that make up the image towards a description of the surfaces and their relative layout in depth (Nakayama, He, & Shimojo, 1995). What emerges from that is perceptual organization: the grouping and sorting of the elements that make up the image. That recurrent processing is mediating perceptual organization is also supported by many experimental findings (Lamme, 1995; Lamme & Spekreijse, 2000; Singer 1999; Zipser, Lamme, & Schiller, 1996).

Perceptual organization is essential for understanding visual perception as it appears to us, i.e., phenomenality. An example can clarify this. Consider seeing a face.

We know that when we see a face, neurons in the inferotemporal cortex are selectively activated. This occurs with extremely short latency, and is probably mediated by feedforward connections (Oram & Perrett, 1992). It is also known, however, that these neurons often signal the presence of a face in a feature-invariant way, that is to say, regardless of its size, position, color, details, or identity (Leopold, Bondar, & Giese, 2006; Rolls, 2000). In other words, these neurons mainly categorize the stimulus as being a face, as opposed to, say, a house or a box. That is not how we see the face. In the phenomenal experience of seeing a face, the face-ness (in a categorical sense) is joined by all the other features that constitute the face, such as its color, shape, emotional expression . . . In other words, a key feature of the phenomenality of seeing the face is that features and categories are integrated. This aspect of seeing the face is not captured by the feedforward activation of the face selective cell, but may be perfectly accounted for by recurrent interactions between face selective neurons and neurons that encode shape, color, etc. (Haxby et al., 2001).

The result of recurrent interactions between neurons is twofold. Neurons in lower regions modify their spiking activity so as to reflect the higher level properties. For example, a V1 neuron receiving feedback signals will fire more strongly when it is responding to features that are part of an object (Albright & Stoner, 2002; Lamme, 1995; Zipser et al., 1996). Conversely, higher level neurons start to reflect lower level properties, which enables them to further categorize the stimulus (Jehee, Roelfsema, Deco, Murre, & Lamme, 2007b). For example, face selective cells that initially only categorize stimuli as face vs. nonface will now become selective for the expression or identity of that face (Sugase, Yamane, Ueno, & Kawano, 1999). Recurrent processing thus makes it possible for neurons to signal shared information.

Sharing of information and feature integration are probably even better supported when neurons engage in recurrent interactions that result in synchronous activity. The role of neural synchrony, whether of an oscillatory nature or not, is still controversial, but much evidence supports the idea that synchronous firing is important for feature binding and conscious perception (Engel, Fries, & Singer, 2001; Engel & Singer, 2001; Singer, 1999; Uhlhaas et al., 2009).

Whether recurrent processing also explains other features of conscious percepts, such as inference or competition, is unclear. Models of visual illusions often use a recurrent architecture (Finkel & Edelman, 1989), but feedforward models exist as well. Cells in V2 respond with very short latency to illusory contours, indicating that at least some inferences are computed in a feedforward fashion in the brain (Peterhans & Vonderheydt, 1989). Likewise, the jury is still out on the exact neural mechanism of perceptual or binocular rivalry (Leopold & Logothetis, 1999; Sterzer, Kleinschmidt, & Rees, 2009). Mutual inhibitory interactions between competing neural assemblies obviously are important, but these may operate at various levels of the visual system (Tong, Meng, & Blake, 2006). An important characteristic of rivalry is that different perceptual levels often switch simultaneously, as do separate regions with similar features. This is best explained by viewing rivalry as a competition between two recurrent assemblies, each representing a single potential percept (Grossberg, Yazdanbakhsh, Cao, & Swaminathan, 2008). Complete proof of the point would be obtained if it could be shown that a dominating percept "kills" the recurrency of the suppressed one.

In sum, recurrent processing has high explanatory power in accounting for important features of conscious percepts, as there is a strong homology between the integrated structure of perception and the structure of recurrent processing.

THE Φ ARGUMENT

Theoretical arguments further stress the importance of recurrent processing in explaining consciousness. Tononi argues that we should understand consciousness as the integration of information (Tononi, 2004, 2008). He uses a measure, Φ, to denote the amount of integrated information that is generated by a system when it goes from one state of processing to the next. This is determined by two factors: First, effective information must be generated, in the sense that the current state of the network—given its way of information processing—rules out a (large) number of previous states. Second, this information must be integrated, so that the amount of information that is generated by the system as a whole is larger than that of the sum of its parts. Tononi uses a powerful metaphor to explain this: Consider a digital camera. Each of the pixels of its sensor carries a bit of information, and so the camera can enter a huge number of different states. The camera is not conscious, however, because it would make no difference when the sensor is divided into individual pixels that work independently. The thalamocortical system consists of elements (neurons or maybe clusters of neurons) that each carry independent bits of information, while at the same time these elements are highly interconnected. That gives it the propensity to generate integrated information, and hence consciousness.

There is a fine balance between the requirements for independent information carried by each element and the elements being interconnected. Too low a connectivity lowers Φ because of the lack of shared information. But too high a connectivity also lowers it, because the elements lose specificity, resulting in less effective information being generated. This aspect of the theory has high explanatory power, in that it explains why the cortico-thalamic network has the capacity to generate high levels of Φ, while other brain structures, such as the cerebellum or basal ganglia, do not (Tononi, 2004). It also explains why we lose consciousness in sleep or epileptic seizures, even while synchronous activity and recurrent processing remain present: Connectivity becomes too high and aspecific (Alkire, Hudetz, & Tononi, 2008; Tononi & Massimini, 2008). Recurrent interactions that produce consciousness therefore should not be too strong (adding to arguments already made by Crick & Koch, 1998b, and countering those by Macknik & Martinez-Conde, 2009). A V1 cell that signals the orientation of a line segment should remain selective for that feature, regardless of whether that feature is part of an object or of the background. But it should alter its response somewhat, reflecting the different context. That is precisely what V1 neurons do in awake animals that report seeing the object (Zipser et al., 1996), and what they don't do during anesthesia (Lamme et al., 1998b), when a percept is absent (Super et al., 2001), or when recurrent interactions are disrupted (Lamme et al., 2002).

Tononi tested different architectures, and feedforward networks typically generate low Φ (Tononi 2004), which explains the absence of consciousness in Stage 1 and Stage 2 processing. In those cases, the brain works more or less like in the digital camera metaphor. Stage 4, on the other hand, probably has high Φ, given the simultaneous specificity and interdependence that is mediated by the recurrent interactions. The presence of consciousness in Stage 4 is therefore readily explained by this theory. Φ is difficult to measure in real networks such as the brain, but related measures are available, such as *causal density* (Seth et al., 2006, 2008). This method uses Granger causality, i.e., the way in which past activity at one point x of a network accounts for the activity at another point y above and beyond past activity of y itself. Like Φ, causal density increases when elements are independently influencing each other. Gaillard et al. (2009) used masking to compare causal density between (roughly) Stage 1/2 and Stage 4 processing, and indeed showed that there is a large difference between the two, confirming these theoretical notions.

The key question here is whether there is a sufficient level of Φ in Stage 3 to grant it phenomenality as well. No direct experimental evidence is available. Theoretically, there probably is a much larger jump in Φ or causal density going from Stage 2 to Stage 3 than going from Stage 3 to Stage 4 (Seth, 2009; Seth et al., 2008). The Stage 2 to Stage 3 transition is characterized by the involvement of horizontal and feedback connections that introduce precisely the interactions that are necessary for high Φ. From Stage 3 to Stage 4, these interactions only grow more widespread. Importantly, Tononi's proposal allows for multiple complexes to coexist at the same time, each supporting its own conscious experience (Tononi, 2004). Therefore, the presence of a Stage 4 complex (the attended objects in Figure 5) does not preclude the presence of a Stage 3 complex (the unattended ones in Figure 5) that is conscious as well.

A FUNDAMENTAL NEURAL DIFFERENCE

It is also relevant to consider whether there are fundamental differences between Stages 1/2 and Stages 3/4—i.e., between feedforward and recurrent processing—other than the ones already mentioned. One such difference could be the extent to which these stages evoke synaptic plasticity. In recurrent processing, in particular when it involves synchronous firing, large numbers of neurons are simultaneously active, satisfying the Hebb rule (Singer, 1995). This is an ideal situation for the massive activation of NMDA receptors and ensuing synaptic plasticity (Dudai, 2002). NMDA receptor activation will obviously also occur during feedforward activation, particularly in the form of spike-time dependent plasticity (Dan & Poo, 2004), but it has been argued that this type of plasticity is all the more effective and specific in the case of oscillatory and synchronous activity in the gamma range, which depends on re-entrant connections. Moreover, for stimulus-specific learning, attention, and consciousness, large scale resonance in thalamocortical circuits combined with spike time dependent plasticity seems essential (Grossberg & Versace, 2008).

These theoretical considerations would imply that learning occurs mainly when neurons engage in recurrent interactions. Consequently, there might be a large jump going from Stage 2 to Stage 3 processing, in the extent to which the different types of processing evoke changes to the brain. Stages 1 and 2 do not evoke much learning, while Stages 3 and 4 do (Lamme, 2006). Indeed, masked stimuli (Stage 2) have only very brief effects on consequent behavior (Eimer & Schlaghecken, 2003), while unattended ones (Stage 3) can have much longer effects, which

are often equal to, and sometimes even larger than those of attended stimuli (Stage 4) (Bornstein, 1989).

Whether indeed recurrent processing is more tightly linked to NMDA receptor activation and learning than feedforward processing still awaits direct experimental evidence (but see Dudkin, Kruchinin, & Chueva, 2001). We have recently obtained evidence that in monkey visual cortex, blockade of NMDA receptors reduces recurrent signals, while blockade of AMPA receptors has its main effect on feedforward activity (unpublished data). Further support comes from studies of anesthesia. It has been noted that many anesthetic agents have as their final common pathway the blockade of NMDA receptor activation (Flohr, Glade, & Motzko, 1998), while at the same time it has been shown that anesthesia abolishes recurrent processing in the visual cortex (Lamme et al., 1998b), leaving feedforward signals relatively untouched.

How would this constitute an argument to grant phenomenality to Stage 3? First of all, it would show once more that there is no fundamental difference between Stages 3 and 4, but only between Stages 3 and 2. However, the argument comes from a neuroscience perspective entirely (Lamme, 2006), which makes it different from the previous ones, which are mixes of phenomenological, functional, and theoretical arguments. Second, the reasoning would improve the ontological status of consciousness, as it can be associated with a basic neural mechanism. This provides a metaphysical argument to grant phenomenality to Stage 3: If we can improve our science of consciousness by granting phenomenality to Stage 3, we—as proper scientists—are forced to do so (Lamme, 2006).

ONTOLOGICAL ISSUES

That brings another player to the stage, which is the science of consciousness. As is readily evident from Figure 4b, by laying the unconscious and conscious processing divide between Stages 1/2 and 3/4 respectively, attention and consciousness (in the sense of phenomenality) become orthogonal and hence independent properties (Lamme, 2003, 2004). There is currently much debate about whether these two functions are better considered independent, and a variety of experiments support this idea (Koch & Tsuchiya, 2007).

In fact, in the same stroke we can make consciousness orthogonal to other functions, such as cognitive control (Figure 4c). For example, both in the Go-NoGo and in the Stop-Signal paradigms, a reaction to a stimulus is required from the subject, unless this stimulus is preceded or followed by a NoGo or Stop stimulus. Such inhibition of planned motor responses typically requires the activation of a prefrontal inhibition network (Eagle et al., 2008; Ridderinkhof, Ullsperger, Crone, & Nieuwenhuiss, 2004). Until recently, it was assumed that this network is only activated by conscious NoGo or Stop signals. It was shown that the very same network is also activated by masked and hence invisible NoGo or Stop signals. This activation furthermore is functional, in that it slows down the response to the visible stimuli (van Gaal et al., 2008, 2009). The finding is readily explained by the schema of Figure 4c. Masked stimuli activate the prefrontal inhibition network via the feedforward sweep (Stage 2), yet remain unconscious because of the absence of recurrent processing. In Stage 4, this activation would have been consciously reportable. But in both cases the function of cognitive control (response inhibition in this case) is executed. Stage 2 yields unconscious cognitive control; Stage 4 yields conscious cognitive control. Of course there are more differences than the phenomenal experience of the inhibitory stimulus between Stages 2 and 4. But these relate mostly to what could be called "motor" phenomenality, such as the sense and strength of control (Rowe, Friston, Frackowiak, & Passingham, 2002).

The schema of Figure 4 not only dissociates consciousness from attention and cognitive control, but may also explain why we can both consciously and unconsciously categorize stimuli (Reder, Park, & Kieffaber, 2009), or even perform unconscious reasoning and inference (Bechara et al., 1997; Dijksterhuis et al., 2006). Which cognitive function is executed depends on *which* area—which cognitive module—is activated. Whether this produces a conscious sensation or not is determined by whether the areas involved engage in recurrent interactions.

The schema of Figure 4, where the conscious–unconscious divide is between Stages 1/2 and Stages 3/4 respectively (Lamme, 2003, 2006), gives a much better ontological status to consciousness (or phenomenality) than Global Neural Workspace (GNW) theory, where that divide lies between Stages 1/2/3 and 4 respectively (Dehaene et al., 2006). In the latter situation, there is no orthogonality between attention and control on one hand and consciousness on the other. In fact, the two are highly confounded. That is another metaphysical argument to grant consciousness to Stage 3. If we only grant phenomenality to Stage 4, we force ourselves into a science of consciousness that is critically flawed from an ontological point of view. The first thing in science always is to demarcate concepts as clearly as possible.

WHY NOT BOTH?

At this point, it will be clear that there are several scientific arguments to conclude that the reason why we have conscious *visual* sensations in Stage 4 is the recurrent interactions between visual areas. These sensations can be reported and manipulated because Stage 4 also includes the prefrontal and motor areas. But if they are not, as in Stage 3, the visual sensation should still be there. All the neural ingredients that seem to matter for visual phenomenality are present in Stage 3, as they are in Stage 4.

From the neural description, we understand why there is access in Stage 4, and not in Stage 3. Stage 4 allows for the widespread integration of visual information with other sensory modalities, motor programs, executive control, and report, simply because here the visual activity is linked with cortical structures that enable such functions (Dehaene et al., 2006). RP in Stage 3 is limited to visual areas, and hence cannot directly influence motor control and other functions necessary for direct report. Several experiments, indeed, have shown that this is what happens in conditions such neglect (Driver & Mattingley, 1998), inattentional blindness (Scholte et al., 2006), or change blindness (Landman et al., 2003b, 2004a; Schankin & Wascher, 2007). In other words, it is perfectly understandable why we have reportable conscious visual sensations in Stage 4, or cognitive access to visual information. There are however no reasons whatsoever to assume that taking away the modules that enable access and report (Stage 3) also takes away the visual phenomenality.

In fact, linking visual phenomenality to access and report gives the whole notion of consciousness a poor ontological status. Stating that consciousness requires both recurrent processing *and* the inclusion of frontal areas (as in GNW theory) seems justified from a behavioral or maybe even an introspective point of view (although the latter is in fact neutral). A closer inspection shows that holding on to this idea impedes progress in our science of consciousness. It disregards the neuroscience argument (Block, 2007; Lamme, 2006). GNW theory is great to explain access, not to explain seeing.

VISION IS RICH EVEN WHEN YOU DON'T KNOW IT

Our conclusion is that Stage 3 processing is just as visually conscious as Stage 4. Change blindness is not blindness, it is the overwriting of one rich conscious visual sensation with another one. You might not know it, but you see all the objects in Figure 1. Likewise, in inattentional blindness, you don't remember having seen an object, but that does not imply you had no conscious visual sensation at the moment it passed by. The event was just not stored in working memory. That is not blindness; it is forgetfulness (Wolfe, 1999). This "seeing without knowing" may sound strange, but in attentional blink paradigms this can be observed directly. When a T1 target is detected, this precludes detection of T2 a few hundred milliseconds later. You really don't know T2 was there. But this only works when the sensation of T2 is wiped out by a following stimulus. A brief presentation of T2, not followed by a mask, gets rid of the attentional blink. Likewise, we should be inclined to conclude that in neglect or extinction, or in split brain patients (Gazzaniga, 2005; Sperry, 1984) the problem is with access and report, but not with seeing.

That is the point of view of a science that goes beyond neural correlates of things we believe to exist introspectively or behaviorally. In this account, neuroscience is used to produce *explanatory correlates* (Seth, 2009) to arrive at a framework with maximal explanatory power regarding consciousness and its relation to other cognitive functions. The approach can also be compared to a factor analysis where behavioral and neural data are simultaneously reduced to underlying principal components, or basic constructs. In psychology, "raw" behavioral data are traditionally boiled down to arrive at underlying concepts such as attention or control. In much cognitive neuroscience, it is then tried to link these concepts to neural structures or mechanisms (Kosslyn, 1999). Here, the behavioral and neural data are taken *together* to arrive at concepts that are better than the ones that can be arrived at by either psychology or neuroscience independently (Lamme, 2004, 2006; Seth, 2008; Seth et al., 2008). In doing so, we automatically move away from behavioral or introspective starting points. If we didn't, neuroscience would not add anything.

This approach is particularly discomforting for consciousness. How can it be that we are mistaken about the identity of a phenomenon that derives its existence from introspection? Is this not a return to behaviorism, with its deep mistrust of anything mental? Or is it a form of eliminative materialism, where mental phenomena are entirely replaced by neural mechanisms, up to the point where talking about the mental would no longer make scientific sense? I think not. I accept the introspective intuition that there is something real and scientifically tractable about conscious experience. In fact it is the starting point of the approach to grant different mental states to Stages 1 and 4—almost entirely from introspection—and from there to extrapolate to Stages 2 and 3. The approach thus is obviously different from behaviorism, maybe even more so than the Global Workspace account, where behavior is taken as the primary evidence for the presence or absence of conscious experience. Neither is there a hidden agenda of eliminative materialism, because I embrace rather than deny the existence of qualia (Dennett, 1988), albeit not in their strictest form. My main objection is against a form of cognitive psychology where mental constructs are taken as undeniable truths to which neuroscience has to be fitted. I would argue that in the study of consciousness, there are no undeniable truths.

That is the standard approach in science. Intuition told us the sun revolves around the earth, while in fact it is the other way around. Intuition dictated creation, where evolution is the counterintuitive scientific answer. To make scientific headway in our science of consciousness, we need to acknowledge that our intuitions may be wrong and need to be set aside. The upshot is that—finally—we may start solving the questions that have been bothering us for the ages.

Commentaries

Stage 3 and what we see

Gideon Paul Caplovitz, Michael J. Arcaro, and Sabine Kastner

Department of Psychology, Princeton University, Princeton, NJ, USA

E-mail: gcaplovi@princeton.edu

DOI: 10.1080/17588928.2010.497584

Abstract: In his article, Lamme provides a neurotheoretical argument that recurrent processing (RP) produces the phenomenological sensations that form the contents of our conscious experiences. Importantly, he argues that this processing includes local intra-areal (i.e., horizontal connections) as well as local inter-areal feedback (i.e., from higher level sensory areas to lower level ones) interactions that occur within the sensory cortices. This has direct implications for what the contents of these experiences may be and the role that neuroscience can play in identifying them.

Lamme argues that we subjectively experience Stage 3 processing that includes local intra-areal (i.e., horizontal connections) as well as local inter-areal feedback (i.e., from higher level sensory areas to lower level ones) interactions that occur within the sensory cortices. Having attributed phenomenology to stage 3, a question remains of how much information it can represent, and how much it thereby contributes to our phenomenological experiences.

Referencing the large capacity representation revealed through studies of iconic memory, Lamme suggests that Stage 3 processing represents a virtual copy of the physical world, and concludes that when examining his Figure 1, one in fact experienced all of the objects in the circular array. This implies Stage 3 processing and that our phenomenological experiences have an unlimited capacity to represent the world around us. However, the capacity of Stage 3 processing (in fact all processing of visual information) is limited by the optics of the eye, the receptive field properties of individual neurons, and their impacts on the dynamics of the circuits to which they belong. Starting at the earliest stages of processing, these physiological limitations lead to numerous ambiguities (e.g., the aperture problem: Adelson & Movshon, 1982) such that an infinite number of physical stimuli can produce the same neuronal activity. It could be argued that a fundamental goal of visual processing in general is to resolve such ambiguities.

As demonstrated by visual illusions and the principles of perceptual grouping, it is clear that we do not experience a direct representation of the physical world, but rather a limited "best guess" of what it might be (Wertheimer, 1924/1950). Certain aspects of how this best guess is constructed, such as the local interactions that support perceptual grouping (i.e., colinear facilitation: Polat & Sagi, 1993), texture segmentation (De Weerd, Sprague, Vandenbussche, & Orban, 1994), as well as competitive interactions that can weaken the representations of co-occurring stimuli (Beck & Kastner, 2009; Desimone & Duncan, 1995), are likely embodied at least partially if not entirely in Stage 3 processing. As such, Stage 3 processing does not allow us to experience a virtual copy of the physical world nor even a direct representation of the retinal image. You likely did not, as Lamme suggests, really see all the objects in his Figure 1, but instead saw the limited information about those objects (as well as the entire visual field) that the visual system was capable of representing.

This leads to an intriguing argument that the contents of our phenomenological experiences should not be considered in terms of the physical stimulus alone, but rather in terms of how information is represented in visual cortex, which, depending on the circumstance, may only loosely correspond to the stimulus itself or even the stimulus-driven inputs. This view provides a direct link between the neuroscience of consciousness and the systems neuroscience of vision. Namely, if we can understand the nature of the information being represented in the visual system, we may begin to understand the contents of our conscious experiences.

Finally, the various parallel pathways (both spatially specific and feature-specific) of visual cortex represent a dynamic system in which activity (both feedforward and recurrent) is ongoing, ever-changing, and dependent not only on its own internal dynamics (e.g., adaptation) but also on numerous inputs that are both stimulus and nonstimulus driven. As such, our phenomenological experiences (and their underlying neural mechanisms) are likely not isolated constructs that appear when a stimulus is present and then disappear when it is removed (as tacitly implied by Lamme's four-stage model), but rather are ever-present and simply change in response to changes that occur within the underlying neural mechanisms. When a stimulus is taken away, there is not a *loss* of consciousness, but a *change* in consciousness. Such changes need not be stimulus driven *per se*, but could arise, for example, from strictly top-down influences of spatial attention (Kastner, Pinsk, De Weerd, Desimone, & Ungerleider, 1999), prior experience (e.g., Bar, 2009), or as a result of internal dynamics as is the case in phenomena such as Troxler filling-in (De Weerd, Gattass, Desimone, & Ungerleider, 1995; Troxler, 1804).

We suggest that in order to understand the contents of our visual experiences, the focus may be better placed on the neural circuits that underlie them (e.g., those that support RP throughout the visual system) and frame questions in terms of how perturbations in the ongoing activity of these circuits arise. Thus, the question is reframed from "How is the conscious experience of a visual stimulus *created*?" to "How is the ongoing stream of conscious experience (and the underlying neural activity) *changed* and what has it *changed to*?" This viewpoint naturally allows the past states of the underlying network to be taken into account when considering how specific inputs will influence the information that will be currently represented.

Furthermore, it naturally dissociates the phenomenological experience of a given stimulus from the stimulus itself.

ACKNOWLEDGMENTS

This work was supported by grants from NIH (RO1 MH64043, RO1 EY017699, P50 MH-62196, T32 MH065214, T90 DA02276) and NSF (BCS-0633281).

* * *

Consciousness minus retrospective mental time travel

Álvaro Machado Dias[1] and Luiz R. G. Britto[2]

[1]*University of São Paulo, Institute of Psychiatry, Neuroimage Lab Rua Dr. Ovidio Pires Campos s/n, São Paulo, Brazil*
[2]*Biomedical Institute, University of São Paulo*
E-mail: alvaromd@usp.br

DOI: 10.1080/17588928.2010.497583

Abstract: In this paper we apply the concept of mental time travel to introduce the basic features of full-blown conscious experiences (encapsulation in mental models and recollection). We discuss the perspective that Lamme's 'Level 3' experiences can be considered as part of the scope of phenomenological consciousness, in relation to which we emphasize the necessity to consider the different degrees of consciousness and how a particular situation compares to the conscious experiences present in resting states of wakefulness.

The understanding of consciousness is intrinsically bound to the notion of time in the mind, much like a DNA double helix. Much as we can gain insights about time by assuming some features about consciousness, the inverse path can aid our understanding of the deep and complex discussion conducted by Lamme.

"Phenomenological consciousness" can serve as a strategic tool in establishing the epistemological basis of a monist philosophical system (as in Merleau-Ponty, 1962, where the anti-dichotomous character is at the heart of the concept), or it can be used simply to express the way we relate to mental content and the outside world. In the latter case, conscious experiences can be thought of as: (1) what (and no more than what) is manifested from instant t to the very next instant (which cannot be defined precisely, for obvious reasons); (2) what (and no less than what) can be framed in mental models and presented to oneself ("access") or to someone else ("reportability").

All levels of reportability relate to past experiences, which are sometimes so close that we do not even realize that we are recollecting. Things are not so clear in relation to conscious access, but the generally accepted idea that full-blown conscious experiences (Lamme's "Level 4") are structurally built (from the bottom up) and then "accessed" (top-down) within specific cognitive domains suggests that access also relates to some sort of past.

The basic process involved in reportability and access (in an even narrower time frame) is the capacity to switch from a modus operandi where construction is in the core (bottom-up) to another that is driven by recollection. In both cases the mechanism may be conceived as a specific type of retrospective mental time travel (MTT) in narrow time frames (Arzy, Adi-Japha, & Blanke, 2009), which encapsulated appearance; conversely, the whole discussion about phenomenality and its relation to particularities of cognitive architecture (e.g., recurrent processing) can be assumed to be the effort to determine whether any level of retrospective MTT is needed for one to consider that some conscious phenomenon has taken place. The idea that someone suffering from a neglect syndrome might have conscious experiences of events that cannot be remembered instants later (see Lamme) can be thought of in terms of an inability to activate MTT within the narrowest intervals.

With this perspective in mind, the boldest and most interesting aspect of Lamme's thesis is his persuasive defense of the standpoint that neglect and other conditions where retrospective MTT is precluded should be considered within the scope of phenomenological consciousness, since the basic biological dimensions of conscious experience are present. However, it is attractive to consider that when we take account of conscious experience in the absence of retrospective MTT, we immediately assume that what is being approached is a manifestation that should be attributed to alterations in parameters usually characterized by low encapsulation and reportability. *Ceteris paribus*, changes in bodily expressions that lead to reportability in the appropriate biocomputational environment (leading to retrospective MTT) can be assumed as indices of conscious experience in Lamme's "Level 3" conditions, since these are equivalent to the former minus encapsulation and activation of the recollection mode.

In effect, this means that it is possible to leave aside the delicate issue of cognitive architecture (recurrent networks vs. synchronicity) and still reach the same endpoint in the debate. For example, orientation reflex (OR) is a fast response that tends to occur before any possibility of conscious access and reportability, which is associated to evoked potentials and changes in skin conductance response (SCR) (Barry & Rushby, 2006); considering that the latter is an index of reportable stress and arousal (Critchley, Elliott, Mathias, & Dolan, 2000), it can be said that their manifestation in, e.g., neglect syndrome for aversive stimuli raises "Level 3" conditions to the status of part of the scope of phenomenological consciousness.

What is most interesting in this example is that the discussion surrounding OR is much more prosaic than the one that relates to neglect syndromes, providing us with an opportunity to consider that the real issue is not whether OR is conscious or not, but the extent to which it is—a question that we can literally think about from minute to minute.

This perspective discloses the importance of integrating the categorical "stage system" to a gradual frame of conscious experience (probably defined by fuzzy parameters). There is no definitive objection to assuming that there is an angle from which all "Level 3" phenomena can be taken as part of the scope of consciousness (regardless of what happens inside the brain). Nevertheless, when we consider the nature and extent of what a subject may be conscious of in these situations where no encapsulation/recollection is provided, we are forced to accept the conclusion that conscious experience in these cases should not be considered to be much different from conscious experience in resting states of wakefulness, which is a practical way of saying that, in effect, it is conscious of nothing except the passage of time from one moment to another.

ACKNOWLEDGMENTS

The first author received grants from The State of São Paulo Research Foundation (FAPESP).

* * *

Explaining seeing? Disentangling qualia from perceptual organization

Agustin Ibáñez[1] and Tristan Bekinschtein[2]

[1]*Institute of Cognitive Neurology (INECO), Buenos Aires, Argentina, and; Career of the National Scientific and Technical Research Council (CONICET), Buenos Aires; Favaloro University, Buenos Aires; and Universidad Diego Portales, Santiago, Chile*
[2]*Medical Research Council, Cambridge, UK, and INECO, Buenos Aires, Argentina*
E-mail: neurologiacognitiva.org

DOI: 10.1080/17588928.2010.497581

Abstract: Visual perception and integration seem to play an essential role in our conscious phenomenology. Relatively local neural processing of reentrant nature may explain several visual integration processes (feature binding or figure–ground segregation, object recognition, inference, competition), even without attention or cognitive control. Based on the above statements, should the neural signatures of visual integration (via reentrant process) be non-reportable phenomenological qualia? We argue that qualia are not required to understand this perceptual organization.

The main point in Lamme's paper is that "we need other evidence about the presence or absence of a conscious sensation than introspection or behavior." (p. 240). We could not be more supportive of this proposal since, in fact, we have been developing measures to address it using electrophysiology (Bekinschtein et al., 2009a) and electromyography (Bekinschtein et al., 2009b). Lamme advocates that seeing is rich but reporting what you see is poor (because the transfer from visual sensation to cognition is limited). Moreover, seeing without access (and report) should have an independent phenomenological quality.

By considering that consciousness is not about access (as proposed by the global neuronal workspace (GNW) theory) but is about phenomenal qualities, Lamme is steering towards visual integration of features as a key component of a conscious percept, and considers qualia a necessary component of it. The GNW does not take the phenomenology of the stimuli into account, and in this sense the criticism of the GNW that working memory is needed to produce qualia loses validity. GNW does not "need" qualia for conscious access (Dehaene & Naccache, 2001).

Furthermore, we consider that even in Lamme's proposal, no qualia are explained but some properties of visual integration (e.g., feature binding or figure–ground segregation, object recognition, inference, competition) are. Those properties may already occur when reentrant process is restricted to a few visual areas (Stage 3). Consequently, Lamme assumes that

Stage 3 is the neural ingredient of phenomenology. Nevertheless, no one of those properties (or their combination) constitutes a *quale* (at least in the phenomenological sense; Block, 1990).

Lamme attributes a phenomenological quality to visual neural processing. It might as well be that there is something that it *is like* to retain briefly a certain number of objects (Stage 3 in Lamme). However, to constitute qualia requires being introspectively accessible, and this is not the case, since Lamme's Stage 3 does not reach the status of being reportable.

There is no need to invoke a phenomenological property to explain visual processing. As Lamme advocates, there are many computational models that yield feature grouping and segregation, figure–ground assignment, depth sorting of surfaces, and occlusion (e.g., Grossberg & Pessoa, 1998; Thielscher & Neumann, 2008), or robots with simulated brain-based recurrent process that learn object recognition (Edelman, 2007). Should we grant qualia properties to those computational networks in order to understand perceptual organization? We believe that *quale* as a property is certainly not needed to understand that kind of perceptual organization (Ibáñez, 2007).

Perceptual organization is part of our experience of qualia (and that is an introspective judgment). But perceptual organization at the same time can be reproduced by natural and artificial neural networks that do not exhibit or require phenomenological properties. Why does Lamme grant phenomenological properties to perceptual organization? Does he implicitly assert an introspective judgment when he classifies a non-necessary phenomenological process (e.g., a reentrant neural process of visual integration) as having a phenomenological property (qualia)? In opposition to Lamme's proposal, this argument does not go beyond introspection.

The neural arguments of Lamme resemble an oldfashioned and frequent categorical error in inter-level explanations of mind (Ryle, 1949). Nevertheless, a fruitful but paradoxical consequence of Lamme's proposal is that introspection allows us to think of qualia *as* perceptual organization. After implicitly making this analogy, we can propose that the neural correlates of qualia (e.g. local reentrant processing) may also have the status of a phenomenological property. It is important to note that this is a metonymic explanation: the part—*e.g. perceptual grouping*—for the whole—*qualia*); probably valid, but not causal. Despite Lamme's assertions, at this stage, his model does not go beyond the neural correlates of a consciousness agenda.

ACKNOWLEDGMENTS

Supported by the CONICET career grant to Agustin Ibáñez.

* * *

How consciousness will change our view on neuroscience

Morten Overgaard

CRNU, Hammel Neurorehabilitation and Research Center, Aarhus University, Denmark
E-mail: mortover@rm.dk

DOI: 10.1080/17588928.2010.497585

Abstract: Victor Lamme proposed that the study of consciousness should not be based on introspection. Nevertheless, Lamme understands consciousness as a subjective phenomenon, and introspection as the way in which we acquire knowledge about consciousness. This makes the task to find introspective-free methods to study consciousness difficult. Lamme attempts to make progress by introducing "neural arguments," but fails to show how such arguments are independent of introspective methods which seem necessary in order to decide how any neural process relates to mental phenomena. This commentary paper thus aims to show that our understanding of neural correlates is shaped by introspection.

Lamme has proposed that our scientific investigations of consciousness should not be based on introspection. The argument states that our intuitive feeling of knowing what we are conscious of may be challenged under certain conditions. When perceiving a visual scene like the one in Sperling's (1960) experiments, we may immediately believe we have consciously seen all the presented stimuli, but after a closer introspective examination we may doubt the validity of this belief.

Before returning to this interpretation of Sperling's experiments, let me however introduce my take on the central concepts involved. Consciousness, however hard to define, seems in most recent publications to be defined as subjective experience. Although Lamme does not present a more formal definition, this seems to be his way of using the term as well. Lamme's conception of introspection seems to be a directing of attention towards the contents of consciousness, and thus different from directing it towards external objects. This definition of introspection resembles those of other researchers (e.g., Jack & Roepstorff, 2002; Overgaard & Sørensen, 2004) and is in opposition to others (e.g., Dretske, 1995). According to Lamme, consciousness does not depend on or necessarily lead to introspection, as we are able to consciously perceive and report about scientific findings with different ("better") validity than we can do based on introspection.

This conception, however, does not come for free, but carries certain necessary consequences. For one thing, introspection becomes the *sine qua non* for knowledge about consciousness. We may think and report about conscious contents only by way of introspection—i.e., attending to consciousness. Being a "disbeliever" in the validity of introspection, Lamme distances himself from other researchers endorsing the same conceptions (e.g. Jack & Roepstorff, 2002; Overgaard, 2006), and he places himself in a rather odd position in trying to study consciousness with little trust in the method he must accept as crucial.

Lamme attempts to get out of this dilemma by introducing "neural arguments." Neural arguments, it seems, differ from neural correlates in such a way that they may be used to make conclusions about conscious experiences. Lamme (p. 213) argues that "we have to decide which are the essential qualities . . . that would 'produce' the phenomenality" and then look for conditions where these essential qualities are present. This, then, would be our introspection-free method to decide whether a subject is conscious without having to ask any direct question about it.

Lamme poses "neural arguments" to decide whether "superficial recurrent processing" should be associated with conscious experience, as Lamme thinks "widespread recurrent processing" should. Were we now to believe that "recurrent processing" is so strongly associated with consciousness that the latter never would appear without the former, we would still not have found such an introspection-free method. To arrive at this association, one would have to conduct several experiments correlating recurrent processes with consciousness—using introspecting experimental participants. Consequently, this method would not be independent of introspection but would carry the strengths, weaknesses, and limitations of introspection. Hence, the "neural argument" method can be no stronger than "neural correlates of introspective reports."

The issue is ironically characterized by the introductory example from the Sperling experiment. The method by which Lamme rejects introspective evidence is . . . introspective evidence. It is only on "closer introspective examination" that doubts may be raised about the validity of the initial, introspectively based belief.

Sadly, the attempt to disregard introspection and find oneself fully dependent on has often been seen before. In fact, it seems a necessary logical consequence of any method suggesting an independent objective measure of consciousness.

According to the view presented here, contrary to the title of Lamme's paper, neuroscience on its own should never change anyone's view on consciousness. However, insights into consciousness and the methods of its study (e.g., Sandberg, Timmermans, Overgaard & Cleeremans, in press) would be of great value to neuroscience, and might indeed change our view of it. By introspection, we form the very categories we put to use when analyzing brain activations as correlations of "colour perception" or "happiness." Thus, neural correlates to subjective states are shaped by introspection in the scientific process.

ACKNOWLEDGMENTS

This work was supported by a Starting Grant, European Research Council.

*   *   *

Electrophysiological evidence for phenomenal consciousness

Antti Revonsuo[1] and Mika Koivisto[2]
[1]*Centre for Cognitive Neuroscience, University of Turku, Turku, Finland, and University of Skövde, Sweden*
[2]*University of Turku, Turku, Finland*
E-mail: revonsuo@utu.fi

DOI: 10.1080/17588928.2010.497580

Abstract: Recent evidence from event-related brain potentials (ERPs) lends support to two central theses in Lamme's theory. The earliest ERP correlate of visual consciousness appears over posterior visual cortex around 100–200 ms after stimulus onset. Its scalp topography and time window are consistent with recurrent processing in the visual cortex. This electrophysiological correlate of visual consciousness is mostly independent of later ERPs reflecting selective attention and working memory functions. Overall, the ERP evidence supports the view that phenomenal consciousness of a visual stimulus emerges earlier than access consciousness, and that attention and awareness are served by distinct neural processes.

Event-related brain potentials (ERPs) track stimulus processing with high temporal resolution. Different ERP waves can be distinguished from each other by their typical onset and peak latency, and their scalp topography. Thus, by exploring the ERP correlates of visual consciousness, it may be possible to answer the following three sets of questions.

1. How many distinct or empirically dissociable ERP waveforms correlate with visual consciousness? More specifically do ERPs reveal only *one* correlate of consciousness or *two separable* correlates of consciousness (one for phenomenal, the other for access consciousness)?

2. In what time window do the ERP correlate(s) of consciousness appear and what kind of scalp topography do they show? More specifically, is there any ERP correlate to be found with a temporal evolution and scalp topography consistent with localized recurrent processing in the visual cortex and thus likely to reflect purely phenomenal consciousness? Or are there only ERP correlates of consciousness that are dependent on attention and working memory, and whose latency and scalp topography suggest widespread "global workspace" processing and frontoparietal involvement? If the latter is true, then the ERP evidence would not support the existence of purely phenomenal consciousness, only the access type of consciousness.

3. Are the ERP correlates of consciousness dissociable from the ERP correlates of attention? More specifically, does the ERP evidence support Lamme's proposal that awareness and attention are served by different neural mechanisms?

To answer questions such as the above, we have recently published a series of ERP experiments of visual awareness and attention. We have also reviewed the relevant wider ERP literature on visual awareness (Koivisto & Revonsuo, 2010). From the published evidence, a coherent pattern emerges that promises to answer the above questions. The consistent pattern of results (and our conclusions below) are based on converging evidence from a number of different kinds of experiments using different experimental paradigms (e.g., masking, change blindness, attentional blink). In all the relevant experiments, the idea has been to contrast a condition where a visual stimulus enters consciousness with a condition where it does not, and to study the differences between these two conditions in the event-related responses of the brain.

Overall, three different candidates for ERP correlates of visual consciousness have emerged: an early positive enhancement around 100 ms from stimulus onset (P1) (Pins & ffytche, 2003), a negative difference wave, visual awareness negativity (VAN; Koivisto & Revonsuo, 2003), typically occurring 150–250 ms from stimulus onset, and a late positive wave (LP) (Niedeggen, Wichmann, & Stoerig, 2001), occurring after 300 ms from stimulus onset. Of these, the P1 enhancement has gained the least support as a genuine correlate of consciousness. It has been only occasionally observed, and probably reflects an attentional effect to near-threshold stimuli that are very difficult to distinguish. By contrast, VAN is the most consistently observed ERP correlate of visual consciousness. Its time window (onset invariably after 100 ms, peaking usually between 200 and 300 ms) and its occipito-temporal, posterior scalp topography are perfectly consistent with the localized recurrent processing in the visual cortex suggested in Lamme's model of consciousness. The LP is similar to other ERP waves in the P3 family of ERPs, with a central and widespread scalp topography. P3 is generally thought to reflect the updating of working memory and other higher cognitive functions that involve frontoparietal attentional networks. Thus, the LP is most naturally interpreted as a correlate of access consciousness.

In experiments where visual consciousness and attention have been separately manipulated it has been possible to test whether phenomenal consciousness (as reflected by VAN) is independent of top-down attention. These experiments have revealed that VAN indeed is independent of selective attention and of the scope of attention (global/local) (Koivisto & Revonsuo, 2008; Koivisto, Revonsuo, & Lehtonen, 2006; Koivisto, Revonsuo, & Salminen, 2005): VAN emerges even for nonselected and nontarget stimuli. The effects of attentional selection affect at most the latter part of VAN.

However, spatial attention appears to be a special case. In a recent experiment (Koivisto, Kainulainen, & Revonsuo, 2009) where spatial attention was strongly manipulated, the stimuli that did not receive any spatial attention also did not elicit any VAN. Therefore, according to these results, spatial attention, but not other forms of attention, is necessary for visual consciousness. On the basis of these results it can be suggested that patients suffering from neglect have no phenomenal consciousness of the neglected stimuli, at least insofar as the stimuli are neglected because of absent spatial attention to them. Here our interpretation is in conflict with Lamme's suggestion that neglect patients are phenomenally conscious of the neglected stimuli.

In conclusion, ERPs are a useful tool to test hypotheses concerning visual awareness and attention. The evidence so far mostly supports the main ideas of Lamme's model: that phenomenal consciousness is independent of higher cognition and attention, but with the important exception of spatial attention.

ACKNOWLEDGMENTS

The authors are supported by the Academy of Finland

* * *

Neural theories need to account for, not discount, introspection and behavior

Anil K. Seth and Adam B. Barrett

University of Sussex, Brighton, UK
E-mail: a.k.seth@sussex.ac.uk

DOI: 10.1080/17588928.2010.496533

Abstract: A satisfying neuroscience of consciousness must account for phenomenological properties in terms of neural properties. While pursuing this project may challenge our intuitions about what we are conscious of, evidence from behavior and introspection should not be discounted. All three lines of evidence need to be integrated in order to naturalize phenomenal experience.

The science of consciousness is rapidly maturing, thanks in large part to highly informative work in cognitive neuroscience, exemplified by Lamme and colleagues. In his target article, Lamme argues that consciousness science is best served by allowing neural evidence to trump intuitions and evidence derived from introspection and behavior. It does seem right to allow that phenomenal consciousness could in principle exist in the absence of introspective or behavioral report (assuming one does not *define* consciousness in terms of reportability). The challenge then is epistemological: How can one infer the presence or absence of consciousness without the validation provided by subjective report? Lamme suggests, and we agree, that a useful approach is to consider which properties of neural processing *account for* rather than merely *correlate with* basic features of phenomenal consciousness (i.e., "explanatory correlates"; Seth, 2009). But we are not convinced that such neural evidence should necessarily trump evidence from other sources.

A neural process may be said to account for an aspect of consciousness when there are identifiable

homologies among neural processes, phenomenal properties, and their cognitive and behavioral accompaniments. Lamme focuses on *recurrent processing* (RP), arguing that it can account for a variety of properties of conscious experience, including figure–ground segregation, feature grouping, occlusion, and the like. So does the existence of RP in Lamme's "stage 3" allow the inference of phenomenal consciousness in the absence of subjective report?

A first challenge is that RP is prevalent within the brain, occurring even in states that are generally considered unconscious including general anesthesia, dreamless sleep, and epileptic absence (Arthuis et al., 2009). RP may therefore be necessary but is unlikely to be sufficient for consciousness. Secondly, some of the properties identified by Lamme, plausibly underpinned by RP, may also characterize unconscious processing. For example, studies of hemispatial neglect have shown that preattentive feature grouping can induce illusory contours in the absence of reportable awareness (Vuilleumier, Valenza, & Landis, 2001). Correct interpretation of these results is, however, difficult. On one hand the absence of reportable awareness in neglect does not necessarily exclude phenomenal consciousness (moreover, the inference to illusory contours still relied on a non-introspective form of behavioral report, line bisection). On the other hand, there seems little *a priori* reason to assume that feature grouping and illusory contour induction are uniquely properties of conscious experience. Thus, Vuilleumier's results could indicate (1) nonreportable phenomenal consciousness, assuming that illusory contours uniquely characterize conscious as opposed to nonconscious contents; (2) reportable phenomenal consciousness, assuming a dissociation between introspective and non-introspective behavioral report; or (3) unconscious processing, challenging the association of phenomenal consciousness with grouping processes of the sort needed for illusory contour induction. We have dwelt on this example because it highlights the difficulty of determining via experiment the fact of the matter about phenomenal consciousness.

To do better, we need explanatory correlates that have stronger *a priori* connections with phenomenal consciousness. One example is that every conscious scene is different from every other possible conscious scene (differentiation), yet is experienced as a unified whole (integration). This property (let's call it "dynamical complexity") is central to recent theoretical approaches including Edelman and Tononi's "dynamic core hypothesis" (Edelman, 2003) and Tononi's "information integration theory" (Tononi, 2008). According to these theories, and *contra* Lamme, we consciously see a face as a face not only virtue of the integration of face-specific perceptual features but also because these integrated features are discriminated from a vast repertoire of alternative possibilities: A face is a face to the extent that it is not a house, a car, the smell of a lemon, an explicit belief, etc.

As Lamme recognizes, dynamical complexity can be operationalized via measures such as neural complexity, Φ (phi), and causal density (Seth, Izhikevich, Reeke, & Edelman, 2006). These measures offer an advance over RP because they are more plausibly associated with conscious as opposed to nonconscious contents (illustrated by the face example above), and because they explain why unconscious conditions (seizures, anesthesia, sleep) can nonetheless show high levels of RP. Unfortunately, existing measures are extremely hard to apply in practice. For example, the current Φ is well defined only for discrete memory-less dynamical systems (Tononi, 2008) and causal density, while measurable from time series data, depends on assumptions of statistical stationarity. More importantly, the plausibility of dynamical complexity as an explanatory correlate is *not* derived purely from neural evidence, but also from introspection (i.e., what are the key invariant phenomenal features of consciousness?) buttressed by inferences relating to behavior and cognition (i.e., that the function of consciousness, with respect to dynamical complexity, is to provide informative discriminations within a vast possibility space of potential conscious scenes).

In conclusion, a mature neuroscience of consciousness will indeed show how neural processes can account for phenomenal properties. In the limit, such processes may shed light on dissociations between phenomenal consciousness and subjective report. However, reaching this limit will require not only more thoroughly worked out explanatory correlates, but also an improved understanding of the mechanisms underpinning report itself. And even then, a mature theory will need to explain why our introspection appears as it does, and which behavioral and cognitive functions are subserved by consciousness, whether accompanied by report or not.

* * *

Localized phenomenology: A recurrent debate

Murray Shanahan

Department of Computing, Imperial College London, London, UK
E-mail: m.shanahan@imperial.ac.uk

DOI: 10.1080/17588928.2010.501405

Abstract: The neuroscience carried out by Lamme and colleagues is fascinating and important. But his case for localised phenomenology rests on a flawed understanding of rival theories and a misguided view of introspective report.

Lamme's paper is the lastest attempt, following in Block's footsteps, to argue for "phenomenology without access." The topic is fraught with difficulty since the relevant data, deriving from experiments such as Sperling's as well as those of Lamme and colleagues, are hard to interpret without taking a stand on philosophical issues. For Lamme, localized recurrent processing (Stage 3) is sufficient for "phenomenology," and the widespread cortical processing that is the hallmark of "access" (Stage 4) is not necessary. Lamme is surely to be applauded for orienting the debate towards the findings of neuroscience. But his arguments are problematic in a number of ways.

In his criticisms of global workspace theory (GWT), Lamme assumes a neural version according to which prefrontal activation precedes broadcast and is therefore a prerequisite for the conscious condition. But this is a faulty assumption. According to the basic tenets of the theory, global broadcast goes hand in hand with the conscious condition. But the idea that a single brain area is the locus of broadcast does not follow from this. A more plausible view is that broadcast is effected by a brain-wide communications infrastructure realized by the cerebral white matter (Shanahan, 2008, 2010). In other words, it's the network that enables access, not some "module" as in Lamme's caricature of the theory. Or more precisely (because "access" is a vague notion), it's thanks to this network that a localized brain process, or coalition of brain processes, can exercise systemic influence.

So Lamme is wrong to ascribe to GWT the view that "taking away the modules that enable access and report . . . also takes away the visual phenomenality." (p. 219). Recent imaging studies have shown that white matter connectivity presents a hierarchically modular small-world topology with a pronounced connective core comprising multiple hub nodes (Hagmann et al., 2008). Such a topology is ideally suited to the global dissemination of influence and information, and is robust to damage. No lesion of an individual hub node in such a network is sufficient to disable the communications infrastructure that is hypothesized to underpin broadcast.

Lamme's misinterpretation of GWT is also apparent when he notes sceptically that "visual information seemingly needs to *go somewhere* to achieve phenomenality" and claims to "smell a homunculus" in the theory (p. 210). Perhaps the "somewhere" Lamme has in mind is the prefrontal cortex. But GWT does not claim that information has to go "somewhere" for the conscious condition to arise. On the contrary, the very essence of the theory is its claim that, to give rise to the conscious condition, information has to go *everywhere*.

The same misunderstanding guides one of Lamme's key arguments for ascribing "phenomenality" to Stage 3 processing. He asks us to consider what distinguishes (indisputably unconscious) Stage 1 and Stage 2 processing from (indisputably conscious) Stage 4 processing. Having dismissed prefrontal processing alone as a candidate (by implication dismissing his misrepresented version of GWT), Lamme alights on the fact that "the remaining difference between Stage 4 and Stages 1 and 2 is that in the latter there is only feedforward processing, while in Stage 4 (and stage 3) there is recurrent processing." (p. 216). The conclusion that recurrent processing is the vital ingredient naturally follows, along with the ascription of "phenomenality" to Stage 3. But *widespread* recurrent processing, the signature of Stage 4, is more than just recurrent *prefrontal* processing. And according to GWT, properly construed, it is widespread activation, not prefrontal activation, that counts. So Lamme is wrong to claim that recurrent processing is the only candidate for what distinguishes Stages 1 and 2 from Stage 4. Wide-spread (recurrent) processing remains a candidate, and GWT as a consequence is still in the frame.

In short, Lamme's claim that recurrent processing alone is sufficient for the conscious condition, even without widespread activation and the resultant capacity for introspective report, is unconvincing. Moreover, there are profoundly important reasons for taking introspective report as a valid indicator of both consciousness and its absence. Suppose a company develops a drug to relieve pain. It works by alleviating the effects of pain only at Lamme's Stage 4, leaving activation at Stage 3 unaffected. (Of course, Lamme's discussion concerns vision, not pain, but his claim must generalize if it is to be taken seriously.) In clinical trials, patients who take the drug report relief from their pain.

However, suppose the authorities refuse to license the drug (under the influence of Lamme's paper, perhaps). Their justification states: "Despite the patients' introspective reports to the contrary, we must assume that the phenomenology of pain is still present, because recurrent neural processing at Stage 3 is unaffected by the drug. The patients only think

they are not in pain. Thanks to neuroscience, we know better." The point of the story is obvious. The neuroscience itself is not in dispute. What matters is how we characterize its findings in ordinary human terms.

* * *

Experiencing more complexity than we can tell

Bert Timmermans[1], Bert Windey[2], and Axel Cleeremans[2]

[1]*Department of Psychiatry, Neuroimaging Group, University of Cologne, Germany*
[2]*Université Libre de Bruxelles, Brussels, Belgium*
E-mail: bert.timmermans@ulb.ac.be

DOI: 10.1080/17588928.2010.497586

Abstract: The notion of unreportable conscious contents is misguidedly premised on the idea that access necessarily follows phenomenal representation. We suggest instead that conscious experience should be viewed as a constructive, dynamical process that involves representational redescription: The brain continuously and unconsciously performs signal detection on its own representations, so developing an understanding of itself that subtends conscious experience. Cases where phenomenality seems to overflow access are thus illusory and depend on interactions between task instructions and stimulus complexity. We support this perspective through recent evidence suggesting that properly graded, qualitative subjective reports appear to be exhaustive in revealing conscious knowledge.

We agree with two intuitions put forward in the target paper. The first is that we see more than we can tell—phenomenal experience does indeed seem to overflow our ability to report on its contents. The second is that consciousness is a graded rather than a dichotomous phenomenon. We note that such gradedness is not incompatible with nonlinearity (as proposed by the global workspace theory, GWT), of which the sigmoid function is a perfect example.

We disagree, however, that consciousness should be defined purely in neural terms. It simply does not make sense to us to speak of unreportable phenomenal contents. Thus, awareness must be distinguished from mere sensitivity. I can be sensitive to some stimulus yet remain unaware of it. In such a case, it makes little sense to think of this sensitivity as involving *any* sort of phenomenal content. It simply reflects the fact that I can react to the stimulus in certain ways, just as an obviously unconscious motion detector can be properly said to be sensitive to movement without this sensitivity implying awareness in any form. Conscious sensitivity, however, crucially involves phenomenal content. It is this specific difference between sensitivity and awareness that one should be focused on.

Both the GWT and the recurrent processing hypothesis (RPH) defend the idea that recurrent processing is essential for conscious experience. However, only GWT assumes that parieto-frontal activity is necessary to amplify initial posterior activity, causing "ignition." In contrast, the RPH proposes that recurrent processing in visual regions is sufficient for conscious perception.

It strikes us, however, that there is a crucial difference in the characteristics of the stimuli used to support either GWT or RPH. Indeed, studies reporting anterior correlates of consciousness generally use complex stimuli (e.g., Del Cul, Baillet, & Dehaene, 2007), whereas studies reporting posterior correlates of consciousness mostly use very low-level stimuli (e.g., Fahrenfort, Scholte, & Lamme, 2008). Whether or not one is a aware of a given stimulus could thus depend on the region that is involved in processing it, so integrating GWT and RPH by letting the emergence of consciousness in Lamme's Stage 3 or Stage 4 depend on the complexity of the stimulus.

In this respect, recent evidence (Sandberg, Timmermans, Overgaard, & Cleeremans, 2010) suggests that for simple stimuli, introspection is in fact rich, graded, and fairly accurate when properly probed by qualitative graded scales referring to the stimulus (from "No experience" to "A clear experience") rather than through dichotomous (yes/no) reports. Strikingly, Overgaard, Fehl, Mouridsen, Bergholt, and Cleeremans (2008) found that, using this method, blindsight patients reveal (severely degraded) awareness of stimuli presented in their blind field. Thus, such graded reports correlate better with behavior and indicate that above-chance identification is always associated with some awareness, at least for simple stimuli.

Whether a subjective report is exhaustive could thus depend on the extent to which feature integration is necessary to respond appropriately to a stimulus (Timmermans, Sandberg, Cleeremans, & Overgaard, 2010). Conversely, stimulus complexity could lie at the heart of the impression of unreportable phenomenal overflow. Kouider, de Gardelle, Sackur, and Dupoux (2010) recently proposed the partial awareness hypothesis, which holds that rich phenomenality is a "perceptual illusion" brought about by partial bottom-up

information that is accessed at *some* but not *all* representational levels, in combination with prior top-down information at the accessed level. Thus, phenomenal awareness never overflows access in this framework. In this sense, becoming aware of a stimulus does not merely involve filtering and selective amplification of a (overflowing) phenomenal field through attention, but rather the active construction of content based on fragmentary input of complex material, biased by priors.

Our own perspective on these issues begins with the notion that the brain *learns to be conscious* by continuously and unconsciously redescribing its own activity to itself (see Cleeremans, 2008). For such redescriptions to be possible at all, the target first-order representations need to be strong, stable, and distinctive—a condition that is itself only possible through recurrent processing. On this view, thus, phenomenal experience depends on the interaction between sufficiently strong first-order representations and the existence of learned redescriptions (metarepresentations) that reflect the manner in which the target first-order representations are known at some level (i.e., their meaning). There may be many levels of such metarepresentations in the brain. Which end up being active during some information-processing episode will depend on both stimulus complexity and task instructions.

* * *

Is recurrent processing necessary and/or sufficient for consciousness?

Naotsugu Tsuchiya[1] and Jeroen J. A. van Boxtel[2]

[1]*Division of Humanities and Social Sciences, California Institute of Technology, Pasadena, CA, USA; and Tamagawa University, Tamagawa, Japan*
[2]*Division of Biology, California Institute of Technology, Pasadena, CA, USA*
E-mail: *naotsu@klab.caltech.edu*

DOI: 10.1080/17588928.2010.497582

Abstract: While we agree with Lamme's general framework, we are not so convinced by his mapping between psychological concepts with their underlying neuronal mechanisms. Specifically, we doubt if recurrent processing is either necessary or sufficient for consciousness. A gist of a scene may be consciously perceived by purely feedforward, without recurrent, processing. Neurophysiological studies of perceptual suppression show recurrent processing in visual cortex for consciously invisible objects. While the neuronal correlates of attention and consciousness remain to be clarified, we agree with Lamme that these two processes are independent, evinced by our recent demonstration of opposing effects of attention and consciousness.

Lamme's hypothesis powerfully explains several phenomena such as masking and iconic memory. However, other lines of research suggest that recurrent processing (RP) is neither necessary nor sufficient for conscious perception.

First, RP seems unnecessary for conscious gist perception of visual scenes. A gist of an unfamiliar and unexpected natural scene can be extracted with a very brief exposure and rapidly reported (Kirchner & Thorpe, 2006). Even when the iconic trace is masked (and presumably RP is terminated), it can be consciously perceived in the near absence of attention (Li, VanRullen, Koch, & Perona, 2002). This happens before the details of the scene become available. These properties of gist perception suggest that its computation can be performed in a purely feedforward manner.

Second, RP seems insufficient for conscious perception. For example, in perceptual suppression phenomena (Leopold & Logothetis, 1996; Maier et al., 2008), objects evoke the same sustained neuronal firing regardless of their conscious visibility. Assuming that sustained firing is a reflection of RP, as Lamme does in other papers, these findings suggest an insufficiency of RP for conscious visibility.

Perhaps, varying amounts of RP are correlated with different kinds of qualia (e.g., no RP for a gist); feedforward activation in an area that has widespread connectivity with the rest of the brain may be sufficient to produce consciousness because it has a higher possibility to produce larger phi. On the other hand, RP in an area that is not connected with frontal areas, like V1, may not give rise to consciousness.

While consciousness may not be tightly correlated with RP, some forms of attention are, and they promote long-distance coherent activity (Womelsdorf & Fries, 2007). In the cases of sustained invisibility mentioned above, some visual aftereffects are enhanced by attention (Kanai, Tsuchiya, & Verstraten, 2006). It would be interesting to study whether attention enhances processing of objects with sustained invisibility via enhanced RP and/or widespread activation.

Although RP and depth of widespread activation may not map onto consciousness and attention, respectively, we do believe that consciousness and attention are supported by distinctive neuronal mechanisms (Tsuchiya & Koch, 2008) based on two lines of evidence: (1) By classifying percepts based on their relation with attention and consciousness, we find examples of attention without consciousness and consciousness without attention, the latter including gist perception and iconic memory. (2) By independently manipulating attention and consciousness, one can demonstrate the opposite effects of attention and consciousness.

As to the second point, the perception of afterimages is modulated in opposite ways by attention and consciousness. By manipulating the visibility (a proxy for the content of consciousness) of an afterimage inducer, *perceptual invisibility* of afterimage inducers is shown to *reduce* afterimage duration. On the other hand, *attending* to afterimage inducers *reduces* afterimage duration. Recently, we directly demonstrated the opposite effects with a 2 × 2 factorial design, removing any stimulus or task confound (van Boxtel, Tsuchiya, & Koch, 2010). We explain the opposite effects by assuming that attention and consciousness enhance luminance and contrast adaptation to different degrees (Brascamp, van Boxtel, Knapen, & Blake, 2010). It would be difficult to explain the opposite effects of attention and consciousness using the line of reasoning based on RP and extent of activation.

While we agree with Lamme's point that neuroscience should go beyond introspection and that attention and consciousness are independent, evidence from gist perception, perceptual suppression, and afterimages suggests that local RP may not explain consciousness.

* * *

DOI: 10.1080/17588928.2010.502224

Reply to Commentaries

What introspection has to offer, and where its limits lie

Victor A. F. Lamme

University of Amsterdam, Amsterdam, The Netherlands

A proper science of consciousness combines all the available evidence – either coming from introspection, behavior, neuroscience or theory – in such a way that a 'best of all worlds' perspective is attained. Introspection shows us that qualia are all about perceptual organization. Neuroscience can then tell us where and when perceptual organization occurs, and whether this is independent of attention, access or report. Access, no matter in what guise it comes, remains ill-suited to explain where, when and how qualia emerge.

INTROSPECTION AND NEUROSCIENCE AS EQUAL PARTNERS

To begin with, I feel some clarification of my position is needed. Some commentaries—Overgaard; Timmermans, Windey, and Cleeremans; Ibáñez and Bekinschtein; Seth and Barrett—see me as a neuroscience chauvinist disregarding introspection (and behavior) altogether. But fully disregarding introspection in favor of neuroscience is not what I propose. On the contrary, my research agenda starts from the notion that qualia—in some form—exist. I follow the intuition that in deep sleep or strong masking there are no conscious sensations, whereas when someone gives a detailed report about what he sees, there are. There are simply no good reasons to doubt these basic intuitions. Without these, the whole point of looking for a better definition of conscious sensation seems moot.

At some point, however, my embracing of intuition stops. While we can all agree on the extreme ends of consciousness, I note that there is sufficient doubt on where exactly the boundary between the conscious and unconscious should be laid (also in the commentaries). I have previously noted that behaviorally, laying such a boundary is impossible (Lamme, 2006). Here I argue that it is equally difficult to do so introspectively. To me that gives sufficient foundation for a scientific debate on that boundary, and neuroscience should be one of the debaters, alongside with psychology, introspection, theoretical reasoning, etc. These contestants should however be on equal footing. That introspection came first doesn't allow it to set the agenda.

WHY PERCEPTUAL ORGANIZATION AND QUALIA ARE LINKED

What I *do* take from introspection is that qualia are better explained by perceptual organization than by access. This is a choice that apparently not everyone agrees on. Ibáñez and Bekinschtein argue that perceptual organization in itself does not constitute a quale, as perceptual organization is just a function—one that can equally well be implemented in a presumably unconscious robot. This argument obviously leads towards the denial of any phenomenality (*à la* Dennett or Churchland; e.g., Churchland, 1985[1]; Dennett, 1988), and indeed they argue that also global workspace theory (GWT) does not need qualia to explain whatever GWT is explaining. If all functionality is taken out of qualia—as in the original formulation of the hard problem—of course all scientific argumentation ends. My stance on qualia is "soft" in the sense that I do link them to some function, in this case perceptual organization.

This has a long empirical background. Gestalt psychologists have always emphasized the importance of perceptual organization for understanding conscious vision. It is hard to imagine any visual illusion occurring unconsciously. Direct evidence for the link

---

[1] Dennett is opposed to the notion of 'pure' qualia, i.e. phenomenality without any consequence. This is probably what Ibanez and Beckinschtein mean by perceptual organization not "requiring" qualia. Qualia as defined by me, strongly linked to perceptual organization, will probably meet much less opposition from Dennett. Churchland, the "father" of eliminative materialism, is in fact more strongly opposed to the propositional attitudes coming from folk psychology (wanting, believing, etc.) than to qualia.

between visual illusions and consciousness came from the work of Goodale and Milner (1992), who showed that only conscious percepts, and not unconscious reflexes, suffer from visual illusions.

This touches on the difference between what Timmermans, Windey, and Cleeremans call sensation and perception. At some point, the information from the outside world that is registered by our senses is transformed into our interpretation of that world. I have always found the phenomenon of color constancy a nice example of that transformation. When we look at an apple on our table, to us that apple will have the same shade of green, regardless of the time of day. The composition of wavelengths that radiate from the apple may however differ greatly depending on whether it is illuminated by morning or afternoon sunlight. Our photoreceptors pick up different wavelengths; we *see* a constant color. Perceptual organization is at the core of explaining this phenomenon: It is by combining the wavelengths of surrounding objects with that of the apple that our brain distills the perceived color. This link between color constancy and consciousness finds direct support in the empirical observation that unconscious priming goes according to wavelength, whereas conscious priming goes according to perceived color (Breitmeyer, Ro, Ogmen, & Todd, 2007).

As Caplovitz, Arcaro, and Kastner eloquently put forward, this going from the physics of the world towards our interpretation of it is what conscious vision is all about. It is the brain that imposes structure and meaning on the incoming information, and conscious sensations are all about how the structure of perception changes from one moment to the next. That is why perceptual organization is the key to understanding qualia. And that is why the boundary between neural representations that go from unconscious to conscious should be laid at the point where neural representations go from, say, wavelengths to colors. Only then does this boundary inform us about the essence of conscious sensations. Putting the boundary at the location where representations go from colors to the access to colors (or maybe their names, or emotional associations) doesn't explain anything about seeing colors instead of wavelengths.

Seth and Barrett do not endorse this *a priori* link between qualia and perceptual organization, feature grouping, or illusions. They argue that even if these were all present in a patient with neglect, it would still be unknown whether the subject had a phenomenal sensation. They argue to look for qualities of neural representations that may have a stronger *a priori* link with qualia, and suggest that these may be found in the intuition that conscious representations seem to be both unified and differentiating: We consciously see a face not only "by virtue of the integration of face-specific perceptual features but also because these integrated features are discriminated from a vast repertoire of alternative possibilities: A face is a face to the extent that it is not a house, a car, the smell of a lemon, . . . etc." (p. 228).

First, in putting so much emphasis on differentiation they lean somewhat towards the unconscious. Face-selective cells, for example, discriminate between a "vast repertoire of alternative" stimuli even when a conscious sensation is undoubtedly absent, as in masking. It is the *combination* of differentiation and integration that makes representations conscious. To me that seems almost synonymous with what perceptual organization or recurrent processing in the visual cortex does: When face-selective cells engage in recurrent interactions with cells that encode lower level features such as color, the face-selective cells start to express both face and color sensitivity, and thus become capable of discriminating not just a face from a house, but also the face of Person A from that of Person B (Sugase, Yamane, Ueno and Kawano, 1999; Jehee, Roelfsema, Deco, Murre and Lamme, 2007b). Simultaneously, the low-level cells start to discriminate between the color red coming from a face and the color red coming from a house. Together, the ensemble of cells constitutes exactly what I would consider a representation that is integrating and differentiating—or, in my words, where perceptual organization has occurred. Therefore, I think that the *a priori* link between integration–differentiation and qualia that Seth and Barrett propose leads to the same conclusion as mine: that neglect patients in fact do have conscious sensations of their unattended/neglected stimuli.

BITING THE ORTHOGONALITY BULLET

This brings another player to the stage, which is attention. It is telling that in most cases where one would argue for the presence of unreportable conscious sensations, attention has been withdrawn from the stimuli at hand (neglect, attentional blink, change blindness, inattentional blindness). The issue boils down to the question of whether taking away attention just removes the ability to access and report these stimuli, or takes away all phenomenality associated with them. By definition, this is a question that cannot be answered introspectively. What does seem open to scientific investigation is the extent to which neural correlates of attention and phenomenal sensations are independent.

Revonsuo and Koivisto provide convincing ERP data to argue for three stages of processing, roughly corresponding to what I call Stages 1/2, 3, and 4. Stage 1/2 processing, reflected in the P1 potential, has no strong correlation with visibility or awareness. The visual awareness negativity (VAN), corresponding to Stage 3 processing, is invariably correlated with visibility. Moreover, this VAN is largely independent of (nonspatial) attention. Stage 4 is reflected in a P3 like positivity (LP) that is reflecting access and attention. These results therefore strongly argue for the independence of attention and consciousness: or of attention and conscious visibility, to be more precise. Tsuchiya and Van Boxtel take this one step further by showing opposite effects of attention and visibility on the duration of after-images.

There thus seems to be increasing experimental support for the independence of attention and consciousness. Together with the ontological argument put forward in my target paper, this makes one wonder why one should not simply bite the bullet and let consciousness and attention be fully orthogonal functions. In the light of many recent results, the arguments seem to swing in favor of doing so.

The notion of attention shares many components with that of access. Both attention and access evoke the amplification of signals, to the extent that they become more widely available. Attention and access have the same behavioral consequences: A target is detected, reported, or put in working memory. Attention and access depend on virtually identical neural structures: the fronto-parietal network. The experimental evidence and ontological arguments in favor of seeing attention as independent of consciousness would therefore also call for making consciousness orthogonal to access. The burden of proof for not doing so is in the hands of those who want to stick to the conflation of consciousness and access.

RECURRENT PROCESSING IN DIFFERENT GUISES

Shanahan posits a somewhat different stance on access, aligned with his interpretation of GWT, where information is broadcast via multiple network hubs via the white matter structure of the cerebral cortex. As soon as localized processing modules are linked through this network to become globally available, access is realized and consciousness ensues. It may be considered a neuroanatomical equivalent of recurrent processing theory or dynamical complexity theory discussed above. And just like in those theories, the key question then becomes how widespread the dissemina-

tion of information has to be before consciousness is produced. Shanahan thinks that the dissemination has to be widespread, but why that is required is unclear. How widespread does it have to be? If these white matter tracts and hubs link visual information to, say, working memory and report modules, but not to auditory or language modules, are we then conscious or not? In the end, also for this position it becomes necessary to answer the question of what modules should be included for a representation to be called conscious.

Dias and Britto consider the amount of "mental time travel" (MTT) associated with a particular representation as critically relevant to the (graded) amount of consciousness that goes along with it. Their idea adds a separate dimension to the discussion: It is not just about which modules are involved (and whether or not these are recurrently activated), but also whether the representation can be thought of to invoke some sort of MTT. The example they give for a representation without MTT—the orienting reflex—is intriguing, because I would consider that to be a strictly feedforward process, i.e., Stage 2. I would say that as soon as any recurrency is involved (i.e., Stages 3 and 4), some sort of MTT is activated along with it, in the sense that there is a meeting of bottom-up construction with top-down expectations. Another argument to link MTT with recurrent processing lies in the relation between recurrent processing and memory formation, as pointed out in the target paper. I thus see the MTT idea as confirmation of the essential dichotomy lying between Stages 1/2 and Stages 3/4.

MORE LEVELS OF CONSCIOUSNESS

Tsuchiya and Van Boxtel empirically question whether recurrent processing is required for conscious experience. Scene gist, for example, can be reported with only very brief presentations of natural scenes, suggesting that they become available through feedforward processing alone. Representations seem to become available in reverse hierarchy, with global properties reaching awareness before details of a scene. Somewhat related is the observation made by Timmermans, Windey, and Cleeremans that access may be operating at different representational levels, from simple to more complex, but is typically accurate for the level of description that is selected.

Indeed, scene gist can be computed in a feedforward sense. The activation of a set of high-level neurons selective for complex and ecologically relevant features (faces, bodies, scene layout, etc.) will give you exactly that. For this representation to become conscious, recurrent interactions between these representations would be sufficient. This is probably what happens when a complex natural scene is partially masked. The masking precludes the interaction with lower level features, and hence the subject is not aware of the scene in all its detail. So partial consciousness of the gist of a complex scene is not incompatible with the recurrent processing theory. The theory predicts that as soon as you have modules representing different aspects of a scene engage in recurrent interactions, you will have a core of conscious sensation linked to whatever is represented by these modules. Adding more modules—of a lower or higher level—makes the conscious sensation of a richer content, but not more or less conscious.

The thought experiment put forward by Shanahan (about the drug that alleviates pain only at Level 4 and not at Level 3, yet is refused a license because neuroscience shows that patients—despite their denial—still "feel" pain) deep down is also about the levels of consciousness. What can we take away from pain and still genuinely call it pain? Obviously, if we take away a subject's capability to report pain by cutting off his tongue, or making him aphasic, the license should indeed not be granted; pain will still be felt. The pain may lose much of its feel if we take away the fear that goes along with it ("is it cancer?"), or the memories or emotions ("this is the worst pain I ever had"). But

wouldn't it still be pain? I even question whether it is critical to attribute the pain to one's own body. Empathic pain may feel just as awful as pain we suffer ourselves. At some point, by taking away sufficient reactive dispositions, we enter the realm of reducing the pain to the level that we should no longer call it pain. When enough has been stripped away, pain may turn into something categorically different, like an itch. This will, however, invariably go along with a change in the Level 3 representation as well. In fact, it is unclear to what extent the feelings of itching and pain are related and may be carried by the same nerve fibers (Ikoma, Steinhoff, Ständer, Yosipovitch, & Schmelz, 2006). Some theories suggest that the main difference between pain and an itch is in the balance or pattern of activation of different afferent fibers (McMahon and Koltzenburg, 1992). So the difference between feeling a pain or an itch may be very similar to the difference between seeing light of 600 nm as either orange or red. It's all about context, about perceptual organization, about combining information.

It is the same with consciousness. From introspection and psychology consciousness looks very different than from neuroscience or theory. Only by putting all these perspectives in their proper context, and by integrating all their information into a unified framework, will we really see what consciousness is about.

References from the Discussion Paper, the Commentaries, and the Reply

Adelson, E. H., & Movshon, J. A. (1982). Phenomenal coherence of moving visual patterns. *Nature*, *30*, 523–525.

Albright, T. D., & Stoner, G. R. (2002). Contextual influences on visual processing. *Annual Review of Neuroscience*, *25*, 339–379.

Alkire, M. T., Hudetz, A. G., & Tononi, G. (2008). Consciousness and anesthesia. *Science*, *322*, 876–880.

Arthuis, M., Valton, L., Régis, J., Chauvel, P., Wendling, F., Naccache, L., et al. (2009). Impaired consciousness during temporal lobe seizures is related to increased long-distance cortical–subcortical synchronization. *Brain*, *132*(8), 2091–2101.

Arzy, S., Adi-Japha, E., & Blanke, O. (2009). The mental time line: An analogue of the mental number line in the mapping of life events. *Consciousness and Cognition*, *18*(3), 781–785.

Baars, B. J. (2005). Global workspace theory of consciousness: Toward a cognitive neuroscience of human experience. In S. Laureys (Ed.), *Boundaries of consciousness:*

Neurobiology and neuropathology (Vol. 150, pp. 45–53). Amsterdam: Elsevier.

Bar, M. (2009). The proactive brain: Memory for predictions. *Philosophical Transactions of the Royal Society B: Biological Sciences*, *364*, 1235–1243.

Barry, R., & Rushby, J. (2006). An orienting reflex perspective on anteriorization of the P3 of the event-related potential. *Experimental Brain Research*, *173*(3), 539–545.

Bechara, A., Damasio, H., Tranel, D. & Damasio, A. R. (1997). Deciding advantageously before knowing the advantageous strategy. *Science*, *275*, 1293–1295.

Beck, D. M., & Kastner, S. (2009). Top-down and bottom-up mechanisms in biasing competition in the human brain. *Vision Research*, *49*, 1154–1165.

Becker, M. W., Pashler, H., & Anstis, S. M. (2000). The role of iconic memory in change-detection tasks. *Perception*, *29*, 273–286.

Bekinschtein, T. A., Dehaene, S., Rohaut, B., Tadel, F., Cohen, L., & Naccache, L. (2009a). Neural signature of the conscious processing of auditory regularities. *Proceedings of the National Academy of Sciences of te United States of America, 3*, 1672–1677.

Bekinschtein, T. A., Shalom, D. E., Forcato, C., Herrera, M., Coleman, M. R., Manes, F. F., et al. (2009b). Classical conditioning in the vegetative and minimally conscious state. *Nature Neuroscience, 12*, 1343–1349.

Blake, R., & Logothetis, N. K. (2002). Visual competition. *Nature Reviews Neuroscience, 3*, 13–23.

Block, N. (1990). Inverted earth. *Philosophical Perspectives, 4*, 53–79.

Block, N. (2005). Two neural correlates of consciousness. *Trends in Cognitive Sciences, 9*, 46–52.

Block, N. (2007). Consciousness, accessibility, and the mesh between psychology and neuroscience. *Behavioral and Brain Sciences, 30*, 481–548.

Boehler, C. N., Schoenfeld, M. A., Heinze, H. J., & Hopf, J. M. (2008). Rapid recurrent processing gates awareness in primary visual cortex. *Proceedings of the National Academy of Sciences of the United States of America, 105*, 8742–8747.

Bornstein, R. F. (1989). Exposure and affect – Overview and meta-analysis of research, 1968–1987. *Psychological Bulletin, 106*, 265–289.

Brascamp, J. W., van Boxtel, J. J., Knapen, T., & Blake, R. (2010). A dissociation of attention and awareness in phase-sensitive but not phase-insensitive visual channels. *Journal of Cognitive Neuroscience, 22*, 2326–2344. doi:10.1162/jocn.2009.21397

Breitmeyer, B. G., Ro, T., Ogmen, H., & Todd, S. (2007). Unconscious, stimulus-dependent priming and conscious, percept-dependent priming with chromatic stimuli. *Perception & Psychophysics, 69*, 550–557.

Bruno, N., & Franz, V. H. (2009). When is grasping affected by the Muller-Lyer illusion? A quantitative review. *Neuropsychologia, 47*, 1421–1433.

Bullier, J. (2001). Integrated model of visual processing. *Brain Research Reviews, 36*, 96–107.

Camprodon, J. A., Zohary, E., Brodbeck, V., & Pascual-Leone, A. (in press). Two phases of V1 activity for visual recognition of natural images. *Journal of Cognitive Neuroscience.* Advance online publication. Retrieved September 31, 2009. doi:10.1162/jocn. 2009.21253

Chalmers, D. J. (1995). Facing up to the problem of consciousness. *Journal of Consciousness Studies, 2*, 200–219.

Churchland, P. M. (1985). Reduction, qualia and the direct inspection of brain states. *Journal of Philosophy, 82*, 8–28.

Churchland, P. S., & Churchland, P. M. (2002). Neural worlds and real worlds. *Nature Reviews Neuroscience, 3*, 903–907.

Cleeremans, A. (2008). Consciousness: The radical plasticity thesis. *Progress in Brain Research, 168*, 19–33.

Coltheart, M. (1980). Iconic memory and visible persistence. *Perception & Psychophysics, 27*, 183–228.

Corballis, P. M. (2003). Visuospatial processing and the right-hemisphere interpreter. *Brain and Cognition, 53*, 171–176.

Crick, F., & Koch, C. (1998a). Consciousness and neuroscience. *Cerebral Cortex, 8*, 97–107.

Crick, F., & Koch, C. (1998b). Constraints on cortical and thalamic projections: The no-strong-loops hypothesis. *Nature, 391*, 245–250.

Crick, F., & Koch, C. (2003). A framework for consciousness. *Nature Neuroscience, 6*, 119–126.

Critchley, H. D., Elliott, R., Mathias, C. J., & Dolan, R. J. (2000). Neural activity relating to generation and representation of galvanic skin conductance responses: A functional magnetic resonance imaging study. *Journal of Neuroscience, 20*(8), 3033–3040.

Dan, Y., & Poo, M. M. (2004). Spike timing-dependent plasticity of neural circuits. *Neuron, 44*, 23–30.

de Gelder, B., Pourtois, G., & Weiskrantz, L. (2002). Fear recognition in the voice is modulated by unconsciously recognized facial expressions but not by unconsciously recognized affective pictures. *Proceedings of the National Academy of Sciences of the United States of America, 99*, 4121–4126.

Dehaene, S., & Naccache, L. (2001). Towards a cognitive neuroscience of consciousness: Basic evidence and a workspace framework. *Cognition, 79*, 1–37.

Dehaene, S., Changeux, J. P., Naccache, L., Sackur, J., & Sergent, C. (2006). Conscious, preconscious, and subliminal processing: A testable taxonomy. *Trends in Cognitive Sciences, 10*, 204–211.

Dehaene, S., Naccache, L., Le Clec'H, G., Koechlin, E., Mueller, M., Dehaene-Lambertz, G., et al. (1998). Imaging unconscious semantic priming. *Nature, 395*, 597–600.

De Weerd, P., Gattass, R., Desimone, R., & Ungerleider, L. G. (1995). Responses of cells in monkey visual cortex during perceptual filling-in of an artificial scotoma. *Nature, 377*(6551), 731–734.

De Weerd, P., Sprague, J. M., Vandenbussche, E., & Orban, G. A. (1994). Two stages in visual texture segregation: A lesion study in the cat. *Journal of Neuroscience, 14*, 929–948.

Del Cul, A., Baillet, S., & Dehaene, S. (2007). Brain dynamics underlying the nonlinear threshold for access to consciousness. *PLoS Biology, 5*(10), e260.

Del Cul, A., Dehaene, S., Reyes, P., Bravo, E., & Slachevsky, A. (2009). Causal role of prefrontal cortex in the threshold for access to consciousness. *Brain, 132*, 2531–2540.

Dennett, D. (1988) Quining qualia. In A. Marcel and E. Bisiach (Eds.), *Consciousness in modern science.* Oxford, UK: Oxford University Press.

Desimone, R. (1998). Visual attention mediated by biased competition in extrastriate visual cortex. *Philosophical Transactions of the Royal Society of London, Series B: Biological Sciences, 353*, 1245–1255.

Desimone, R., & Duncan, J. (1995). Neural mechanisms of selective visual-attention. *Annual Review of Neuroscience, 18*, 193–222.

Dijksterhuis, A., Bos, M. W., Nordgren, L. F., & van Baaren, R. B. (2006). On making the right choice: The deliberation-without-attention effect. *Science, 311*, 1005–1007.

Dretske, F. (1995). *Naturalizing the mind.* Cambridge, MA: MIT Press.

Driver, J., & Mattingley, J. B. (1998). Parietal neglect and visual awareness. *Nature Neuroscience, 1*, 17–22.

Dudai, Y. (2002). Molecular bases of long-term memories: A question of persistence. *Current Opinion in Neurobiology, 12*, 211–216.

Dudkin, K. N., Kruchinin, V. K., & Chueva, I. V. (2001). Neurophysiological correlates of delayed differentiation tasks in monkeys: The effects of the site of intracortical blockade of NMDA receptors. *Neuroscience and Behavioral Physiology, 31*, 207–218.

Eagle, D. M., Baunez, C., Hutcheson, D. M., Lehmann, O., Shah, A. P., & Robbins, T. W. (2008). Stop-signal reaction-time task performance: Role of prefrontal cortex and subthalamic nucleus. *Cerebral Cortex*, *18*, 178–188.

Edelman, G. M. (1992). *Bright air, brilliant fire: On the matter of the mind.* New York: Basic Books.

Edelman, G. M. (2003). Naturalizing consciousness: A theoretical framework. *Proceedings of the National Academy of Sciences of the United States of America*, *100*(9), 5520–5524.

Edelman, G. M. (2007). Learning in and from brain-based devices. *Science*, *16*(318), 1103–1105.

Egeth, H. E., & Yantis, S. (1997). Visual attention: Control, representation, and time course. *Annual Review of Psychology*, *48*, 269–297.

Eimer, M., & Schlaghecken, F. (2003). Response facilitation and inhibition in subliminal priming. *Biological Psychology*, *64*, 7–26.

Engel, A. K., Fries, P., & Singer, W. (2001). Dynamic predictions: Oscillations and synchrony in top-down processing. *Nature Reviews Neuroscience*, *2*, 704–716.

Engel, A. K., & Singer, W. (2001). Temporal binding and the neural correlates of sensory awareness. *Trends in Cognitive Sciences*, *5*, 16–25.

Enns, J. T., & Di Lollo, V. (2000). What's new in visual masking? *Trends in Cognitive Sciences*, *4*, 345–352.

Fahrenfort, J. J., Scholte, H. S., & Lamme, V. A. F. (2007). Masking disrupts reentrant processing in human visual cortex. *Journal of Cognitive Neuroscience*, *19*, 1488–1497.

Fahrenfort, J. J., Scholte, H. S., & Lamme, V. A. F. (2008). The spatiotemporal profile of cortical processing leading up to visual perception. *Journal of Vision*, *8*(1), 1–12.

Finkel, L. H., & Edelman, G. M. (1989). Integration of distributed cortical systems by reentry: A computer-simulation of interactive functionally segregated visual areas. *Journal of Neuroscience*, *9*, 3188–3208.

Flohr, H., Glade, U., & Motzko, D. (1998). The role of the NMDA synapse in general anesthesia. *Toxicology Letters*, *101*, 23–29.

Gaillard, R., Dehaene, S., Adam, C., Clemenceau, S., Hasboun, D., Baulac, M., et al. (2009). Converging intracranial markers of conscious access. *PLOS Biology*, *7*, 472–492.

Gazzaniga, M. S. (2005). Forty-five years of split-brain research and still going strong. *Nature Reviews Neuroscience*, *6*, 653–659.

Goodale, M. A., & Milner, A. D. (1992). Separate visual pathways for perception and action. *Trends in Neuroscience*, *15*, 20–25.

Grossberg, S., & Pessoa, L. (1998). Texture segregation, surface representation and figure–ground separation. *Vision Research*, *38*, 2675–2684.

Grossberg, S. & Versace, M. (2008). Spikes, synchrony, and attentive learning by laminar thalamocortical circuits. *Brain Research*, *1218*, 278–312.

Grossberg, S., Yazdanbakhsh, A., Cao, Y. Q., & Swaminathan, G. (2008). How does binocular rivalry emerge from cortical mechanisms of 3-D vision? *Vision Research*, *48*, 2232–2250.

Haber, R. N. (1983). The impending demise of the icon: A critique of the concept of iconic storage in visual information-processing. *Behavioral and Brain Sciences*, *6*, 1–11.

Hagmann, P., Cammoun, L., Gigandet, X., Meuli, R., Honey, C. J., Wedeen, C. J., et al. (2008). Mapping the structural core of human cerebral cortex. *PLoS Biology*, *6*(7), e159.

Haxby, J. V., Gobbini, M. I., Furey, M. L., Ishai, A., Schouten, J. L., & Pietrini, P. (2001). Distributed and overlapping representations of faces and objects in ventral temporal cortex. *Science*, *293*, 2425–2430.

Haynes, J. D. (2009). Decoding visual consciousness from human brain signals. *Trends in Cognitive Sciences*, *13*, 194–202.

Haynes, J. D., Driver, J., & Rees, G. (2005). Visibility reflects dynamic changes of effective connectivity between V1 and fusiform cortex. *Neuron*, *46*, 811–821.

Ibáñez A. (2007). The neurodynamic core of consciousness and neural Darwinism. *Revista de Neurologia*, *45*, 547–555.

Ikoma, A., Steinhoff, M., Ständer, S., Yosipovitch, G., & Schmelz, M. (2006). The neurobiology of itch. *Nature Reviews Neuroscience*, *7*, 535–547.

Jack, A., & Roepstorff, A. (2002). Retrospection and cognitive brain mapping: From stimulus-response to script-report. *Trends in Cognitive Sciences*, *6*, 333–339.

Jehee, J. F. M., Lamme, V. A. F., & Roelfsema, P. R. (2007a). Boundary assignment in a recurrent network architecture. *Vision Research*, *47*, 1153–1165.

Jehee, J. F. M., Roelfsema, P. R., Deco, G., Murre, J. M. J., & Lamme, V. A. F. (2007b). Interactions between higher and lower visual areas improve shape selectivity of higher level neurons: Explaining crowding phenomena. *Brain Research*, *1157*, 167–176.

Kanai, R., Tsuchiya, N., & Verstraten, F. A. (2006). The scope and limits of top-down attention in unconscious visual processing. *Current Biology*, *16*(23), 2332–2336.

Kastner, S., Pinsk, M. A., De Weerd, P., Desimone R., & Ungerleider, L. G. (1999). Increased activity in human visual cortex during directed attention in the absence of visual stimulation. *Neuron*, *22*, 751–761.

Kentridge, R. W., Heywood, C. A., & Weiskrantz, L. (1999). Attention without awareness in blindsight. *Proceedings of the Royal Society of London, Series B: Biological Sciences*, *266*, 1805–1811.

Kirchner, H., & Thorpe, S. J. (2006). Ultra-rapid object detection with saccadic eye movements: Visual processing speed revisited. *Vision Research*, *46*(11), 1762–1776.

Klein, T. A., Endrass, T., Kathmann, N., Neumann, J., von Cramon, D. Y., & Ullsperger, M. (2007). Neural correlates of error awareness. *NeuroImage*, *34*, 1774–1781.

Koch, C., & Tsuchiya, N. (2007). Attention and consciousness: Two distinct brain processes. *Trends in Cognitive Sciences*, *11*, 16–22.

Koivisto, M., Kainulainen, P., & Revonsuo, A (2009). The relationship between awareness and attention: evidence from ERP responses. *Neuropsychologia*, *47*, 2891–2899.

Koivisto, M., & Revonsuo, A. (2003). An ERP study of change detection, change blindness and visual awareness. Psychophysiology, 40, 423–429.

Koivisto, M., & Revonsuo, A. (2008). The role of selective attention in visual awareness of stimulus features: Electrophysiological studies. *Cognitive, Affective, & Behavioral Neuroscience*, *8*, 195–210.

Koivisto, M., & Revonsuo, A. (2010). Event-related brain potential correlates of visual awareness. *Neuroscience and Biobehavioral Reviews, 34*, 922–934.

Koivisto, M., Revonsuo, A., & Lehtonen, M. (2006). Independence of visual awareness from the scope of attention: An electrophysiological study. *Cerebral Cortex, 16*, 415–424.

Koivisto, M., Revonsuo, A., & Salminen, N. (2005). Independence of visual awareness from attention at early processing stages. *NeuroReport, 16*, 817–821.

Kosslyn, S. M. (1999). If neuroimaging is the answer, what is the question? *Philosophical Transactions of the Royal Society of London, Series B: Biological Sciences, 354*, 1283–1294.

Kouider, S., de Gardelle, V., Sackur, J., & Dupoux, E. (in press). How rich is consciousness? The partial awareness hypothesis. *Trends in Cognitive Sciences.* Advance online publication. doi:10.1016/j.tics.2010.04.006

Lamme, V. A. F. (1995). The neurophysiology of figure ground segregation in primary visual-cortex. *Journal of Neuroscience, 15*, 1605–1615.

Lamme, V. A. F. (2003). Why visual attention and awareness are different. *Trends in Cognitive Sciences, 7*, 12–18.

Lamme, V. A. F. (2004). Separate neural definitions of visual consciousness and visual attention: A case for phenomenal awareness. *Neural Networks, 17*, 861–872.

Lamme, V. A. F. (2006). Towards a true neural stance on consciousness. *Trends in Cognitive Sciences, 10*, 494–501.

Lamme, V. A. F., & Roelfsema, P. R. (2000). The distinct modes of vision offered by feedforward and recurrent processing. *Trends in Neurosciences, 23*, 571–579.

Lamme, V. A. F., & Spekreijse, H. (2000). Modulations of primary visual cortex activity representing attentive and conscious scene perception. *Frontiers in Bioscience, 5*, D232–D243.

Lamme, V. A. F., Super, H., Landman, R., Roelfsema, P. R. & Spekreijse, H. (2000). The role of primary visual cortex (V1) in visual awareness. *Vision Research, 40*, 1507–1521.

Lamme, V. A. F., Super, H. & Spekreijse, H. (1998a). Feedforward, horizontal, and feedback processing in the visual cortex. *Current Opinion in Neurobiology, 8*, 529–535.

Lamme, V. A. F., Vandijk, B. W., & Spekreijse, H. (1993). Contour from motion processing occurs in primary visual-cortex. *Nature, 363*, 541–543.

Lamme, V. A. F., Zipser, K., & Spekreijse, H. (1998b). Figure–ground activity in primary visual cortex is suppressed by anesthesia. *Proceedings of the National Academy of Sciences of the United States of America, 95*, 3263–3268.

Lamme, V. A. F., Zipser, K., & Spekreijse, H. (2002). Masking interrupts figure–ground signals in V1. *Journal of Cognitive Neuroscience, 14*, 1044–1053.

Landman, R., Spekreijse, H., & Lamme, V. A. F. (2003a). Large capacity storage of integrated objects before change blindness. *Vision Research, 43*, 149–164.

Landman, R., Spekreijse, H., & Lamme, V. A. F. (2003b). Set size effects in the macaque striate cortex. *Journal of Cognitive Neuroscience, 15*, 873–882.

Landman, R., Spekreijse, H. & Lamme, V. A. F. (2004a). Relationship between change detection and post-change activity in visual area VI. *NeuroReport, 15*, 2211–2214.

Landman, R., Spekreijse, H., & Lamme, V. A. F. (2004b). The role of figure–ground segregation in change blindness. *Psychonomic Bulletin & Review, 11*, 254–261.

Lau, H. C. & Passingham, R. E. (2007). Unconscious activation of the cognitive control system in the human prefrontal cortex. *Journal of Neuroscience, 27*, 5805–5811.

Lee, S. H., Blake, R., & Heeger, D. J. (2007). Hierarchy of cortical responses underlying binocular rivalry. *Nature Neuroscience, 10*, 1048–1054.

Leopold, D. A., Bondar, I. V., & Giese, M. A. (2006) Norm-based face encoding by single neurons in the monkey inferotemporal cortex. *Nature, 442*, 572–575.

Leopold, D. A., & Logothetis, N. K. (1996). Activity changes in early visual cortex reflect monkeys' percepts during binocular rivalry. *Nature, 379*(6565), 549–553.

Leopold, D. A., & Logothetis, N. K. (1999). Multistable phenomena: Changing views in perception. *Trends in Cognitive Sciences, 3*, 254–264.

Li, F. F., VanRullen, R., Koch, C., & Perona, P. (2002). Rapid natural scene categorization in the near absence of attention. *Proceedings of the National Academy of Sciences of the United Sstates of America, 99*(14), 9596–9601.

Luck, S. J., & Vogel, E. K. (1997). The capacity of visual working memory for features and conjunctions. *Nature, 390*, 279–281.

Mack, A. & Rock, I. (1998). *Inattentional blindness.* Cambridge, MA: MIT Press.

Macknik, S. L., & Martinez-Conde, S. (2009). The role of feedback in visual attention and awareness. In M. S. Gazzaniga (Ed.), *The cognitive neurosciences* (pp. 1165–1179). Cambridge, MA: MIT Press.

Maier, A., Wilke, M., Aura, C., Zhu, C., Ye, F. Q., & Leopold, D. A. (2008). Divergence of fMRI and neural signals in V1 during perceptual suppression in the awake monkey. *Nature Neuroscience, 11*(10), 1193–1200.

Marois, R., Yi, D. J., & Chun, M. M. (2004). The neural fate of consciously perceived and missed events in the aftentional blink. *Neuron, 41*, 465–472.

McMahon, S. B. and Koltzenburg, M. (1992) Itching for an explanation. *Trends in Neuroscience, 15*, 497–501.

Meng, M., & Tong, F. (2004). Can attention selectively bias bistable perception? Differences between binocular rivalry and ambiguous figures. *Journal of Vision, 4*, 539–551.

Merleau-Ponty, M. (1962). *Phenomenology of perception.* London: Routledge & Kegan Paul.

Nakayama, K., He, Z. J., & Shimojo, S. (1995). Visual surface representation: A critical link between lower-level and higher level vision. In S. M. Kosslyn & D. N. Osherson (Eds.), *An invitation to cognitive science: Visual cognition* (pp. 1–70). Cambridge, MA: MIT Press.

Nakayama, K., & Shimojo, S. (1992). Experiencing and perceiving visual surfaces. *Science, 257*, 1357–1363.

Niedeggen, M., Wichmann, P., & Stoerig, P. (2001). Change blindness and time to consciousness. *European Journal of Neuroscience, 14*, 1719–1726.

O'Regan, J. K., & Noe, A. (2001). A sensorimotor account of vision and visual consciousness. *Behavioral and Brain Sciences, 24*, 939–973.

O'Shea, R. P., & Corballis, P. M. (2003). Binocular rivalry in split-brain observers. *Journal of Vision, 3*, 610–615.

Oram, M. W., & Perrett, D. I. (1992). Time course of neural responses discriminating different views of the face and head. *Journal of Neurophysiology, 68*, 70–84.

Overgaard, M. (2006). Introspection in science. *Consciousness and Cognition, 15*, 629–633.

Overgaard, M., Fehl, K., Mouridsen, K., Bergholt, B., & Cleeremans, A. (2008). Seeing without seeing? Degraded conscious vision in a blindsight patient. *PLoS ONE, 3*(8), e3028.

Overgaard, M., & Sørensen, T. A. (2004). Introspection distinct from first order experiences. *Journal of Consciousness Studies, 11*(7–8), 77–95.

Pascual-Leone, A., & Walsh, V. (2001). Fast backprojections from the motion to the primary visual area necessary for visual awareness. *Science, 292*, 510–512.

Peterhans, E., & Vonderheydt, R. (1989). Mechanisms of contour perception in monkey visual-cortex. 2. Contours bridging gaps. *Journal of Neuroscience, 9*, 1749–1763.

Pins, D., & ffytche, D. (2003). The neural correlates of conscious vision. *Cerebral Cortex, 13*, 461–44.

Polat U., & Sagi D. (1993). Lateral interactions between spatial channels: Suppression and facilitation revealed by lateral masking experiments. *Vision Research, 33*, 993–999.

Pylyshyn, Z. (1999). Is vision continuous with cognition? The case for cognitive impenetrability of visual perception. *Behavioral and Brain Sciences, 22*, 341–365.

Reder, L. M., Park, H., & Kieffaber, P. D. (2009). Memory systems do not divide on consciousness: Reinterpreting memory in terms of activation and binding. *Psychological Bulletin, 135*, 23–49.

Rees, G. (2007). Neural correlates of the contents of visual awareness in humans. *Philosophical Transactions of the Royal Society, B: Biological Sciences, 362*, 877–886.

Rees, G., Kreiman, G., & Koch, C. (2002). Neural correlates of consciousness in humans. *Nature Reviews Neuroscience, 3*, 261–270.

Ridderinkhof, K. R., Ullsperger, M., Crone, E. A., & Nieuwenhuiss, S. (2004). The role of the medial frontal cortex in cognitive control. *Science, 306*, 443–447.

Roelfsema, P. R., Lamme, V. A. F., Spekreijse, H., & Bosch, H. (2002). Figure–ground segregation in a recurrent network architecture. *Journal of Cognitive Neuroscience, 14*, 525–537.

Rolls, E. T. (2000). Functions of the primate temporal lobe cortical visual areas in invariant visual object and face recognition. *Neuron, 27*, 205–218.

Rolls, E. T., & Tovee, M. J. (1994). Processing speed in the cerebral-cortex and the neurophysiology of visual masking. *Proceedings of the Royal Society of London, Series B: Biological Sciences, 257*, 9–15.

Rowe, J., Friston, K., Frackowiak, R., & Passingham, R. (2002). Attention to action: Specific modulation of corticocortical interactions in humans. *NeuroImage, 17*, 988–998.

Ryle, G. (1949). *The concept of mind.* Chicago: New University of Chicago Press.

Salin, P. A., & Bullier, J. (1995). Corticocortical connections in the visual-system: Structure and function. *Physiological Reviews, 75*, 107–154.

Sandberg, K., Timmermans, B., Overgaard, M., & Cleeremans, A. (in press). Measuring consciousness: Is one measure better than the other? *Consciousness and Cognition.* Advance online publication. doi:10.1016/j.concog.2009.12.013

Schacter, D. L., Chiu, C. Y. P., & Ochsner, K. N. (1993). Implicit memory – A selective review. *Annual Review of Neuroscience, 16*, 159–182.

Schankin, A., & Wascher, E. (2007). Electrophysiological correlates of stimulus processing in change blindness. *Experimental Brain Research, 183*, 95–105.

Scholte, H. S., Witteveen, S. C., Spekreijse, H., & Lamme, V. A. F. (2006). The influence of inattention on the neural correlates of scene segmentation. *Brain Research, 1076*, 106–115.

Serences, J. T., & Yantis, S. (2006). Selective visual attention and perceptual coherence. *Trends in Cognitive Sciences, 10*, 38–45.

Sergent, C., Baillet, S., & Dehaene, S. (2005). Timing of the brain events underlying access to consciousness during the attentional blink. *Nature Neuroscience, 8*, 1391–1400.

Seth, A. K. (2008). Theories and measures of consciousness develop together. *Consciousness and Cognition, 17*, 986–988.

Seth, A. K. (2009). Explanatory correlates of consciousness: Theoretical and computational challenges. *Cognitive Computation, 1*(1), 50–63.

Seth, A. K., Dienes, Z., Cleeremans, A., Overgaard, M., & Pessoa, L. (2008). Measuring consciousness: Relating behavioural and neurophysiological approaches. *Trends in Cognitive Sciences, 12*, 314–321.

Seth, A. K., Izhikevich, E., Reeke, G. N., & Edelman, G. M. (2006). Theories and measures of consciousness: An extended framework. *Proceedings of the National Academy of Sciences of the United States of America, 103*(28), 10799–10804.

Shanahan, M. P. (2008). A spiking neuron model of cortical broadcast and competition. *Consciousness and Cognition, 17*, 288–303.

Shanahan, M. P. (2010). *Embodiment and the inner life: Cognition and consciousness in the space of possible minds.* Oxford, UK: Oxford University Press.

Shapiro, K. L. (2009). The functional architecture of divided visual attention. *Progress in Brain Research, 176*, 101–121.

Shapiro, K. L., Raymond, J. E., & Arnell, K. M. (1994). Attention to visual-pattern information produces the attentional blink in rapid serial visual presentation. *Journal of Experimental Psychology: Human Perception and Performance, 20*, 357–371.

Silvanto, J., Cowey, A., Lavie, N., & Walsh, V. (2005). Striate cortex (V1) activity gates awareness of motion. *Nature Neuroscience, 8*, 143–144.

Simons, D. J., & Rensink, R. A. (2005). Change blindness: Past, present, and future. *Trends in Cognitive Sciences, 9*, 16–20.

Singer, W. (1995). Development and plasticity of cortical processing architectures. *Science, 270*, 758–764.

Singer, W. (1999). Neuronal synchrony: A versatile code for the definition of relations? *Neuron, 24*, 49–65.

Sligte, I. G., Scholte, H. S., & Lamme, V. A. F. (2008). Are there multiple visual short-term memory stores? *PLOS One, 3*, e1699.

Sligte, I. G., Scholte, H. S., & Lamme, V. A. F. (2009). V4 activity predicts the strength of visual short-term memory representations. *Journal of Neuroscience, 29*, 7432–7438.

Sperling, G. (1960). The information available in brief visual presentations, *Psychological Monographs, 74*, 1–29.

Sperry, R. (1984). Consciousness, personal identity and the divided brain. *Neuropsychologia, 22*, 661–673.

Sporns, O., Tononi, G., & Edelman, G. M. (1991). Modeling perceptual grouping and figure ground segregation by means of active reentrant connections. *Proceedings of the National Academy of Sciences of the United States of America, 88,* 129–133.

Sterzer, P., Kleinschmidt, A., & Rees, G. (2009). The neural bases of multistable perception. *Trends in Cognitive Sciences, 13,* 310–318.

Stoerig, P. (1996). Varieties of vision: From blind responses to conscious recognition. *Trends in Neurosciences, 19,* 401–406.

Sugase, Y., Yamane, S., Ueno, S., & Kawano, K. (1999). Global and fine information coded by single neurons in the temporal visual cortex. *Nature, 400,* 869–873.

Super, H., Spekreijse, H., & Lamme, V. A. F. (2001). Two distinct modes of sensory processing observed in monkey primary visual cortex (V1). *Nature Neuroscience, 4,* 304–310.

Thielscher, A., & Neumann, H. (2008). Globally consistent depth sorting of overlapping 2D surfaces in a model using local recurrent interactions. *Biological Cybernetics, 98,* 305–337.

Thompson, K. G., & Schall, J. D. (1999). The detection of visual signals by macaque frontal eye field during masking. *Nature Neuroscience, 2,* 283–288.

Timmermans, B., Sandberg, K., Cleeremans, A., & Overgaard, M. (in press). Partial awareness distinguishes between measuring conscious perception and conscious content. *Consciousness and Cognition.* Advance online publication. doi:10.1016/j.concog.2010.05.006

Tong, F., Meng, M., & Blake, R. (2006). Neural bases of binocular rivalry. *Trends in Cognitive Sciences, 10,* 502–511.

Tononi, G. (2004). An information integration theory of consciousness. *BMC Neuroscience, 5,* 42.

Tononi, G. (2008). Consciousness as integrated information: A provisional manifesto. *Biological Bulletin, 215,* 216–242.

Tononi, G., & Koch, C. (2008). The neural correlates of consciousness: An update. In *Year in Cognitive Neuroscience 2008* (vol. 1124, pp. 239–261). Boston: Blackwell.

Tononi, G., & Massimini, M. (2008). Why does consciousness fade in early sleep? In D. W. Pfaff and B. L. Kieffer (Eds.), *Molecular and biophysical mechanisms of arousal, alertness, and attention* (Vol. 1129, pp. 330–334). New York: New York Academy of Sciences.

Treisman, A. (1996). The binding problem. *Current Opinion in Neurobiology, 6,* 171–178.

Troxler, D. (1804). Uber das Verschwinden gegebener Gegenstände innerhalb unseres Gesichtskreises [On the disappearance of given objects in visual space]. In K. Himly & J. A. Schmidt (Eds.), *Ophthalmologische bibliothek II* (pp. 51–53). Jena, Germany: Fromman.

Tsuchiya, N., & Koch, C. (2008). The relationship between consciousness and attention. In S. Laureys & G. Tononi (Eds.), *The neurology of consciousness: Cognitive neuroscience and neuropathology* (pp. 63–78). New York: Academic Press.

Uhlhaas, P. J., Pipa, G., Melloni, L., Neuenschwander, S., Nikolic, D. & Singer, W. (2009). Neural synchrony in cortical networks: History, concept and current status. *Frontiers in Integrative Neuroscience, 3,* 17.

van Boxtel, J. J., Tsuchiya, N., & Koch, C. (2010). Opposing effects of attention and consciousness on afterimages. *Proceedings of the National Academy of Sciences of the United Sstates of America, 107*(19), 8883–8888.

van Gaal, S., Ridderinkhof, K. R., Fahrenfort, J. J., Scholte, H. S., & Lamme, V. A. F. (2008). Frontal cortex mediates unconsciously triggered inhibitory control. *Journal of Neuroscience, 28,* 8053–8062.

van Gaal, S., Ridderinkhof, K. R., van den Wildenberg, W. P. M., & Lamme, V. A. F. (2009). Dissociating consciousness from inhibitory control: Evidence for unconsciously triggered response inhibition in the stop-signal task. *Journal of Experimental Psychology: Human Perception and Performance, 35,* 1129–1139.

Vuilleumier, P., Valenza, N., & Landis, T. (2001). Explicit and implicit perception of illusory contours in unilateral spatial neglect: Behavioural and anatomical correlates of preattentive grouping mechanisms. *Neuropsychologia, 39*(6), 597–610.

Watanabe, T., Nanez, J. E., & Sasaki, Y. (2001). Perceptual learning without perception. *Nature, 413,* 844–848.

Wertheimer M. (1924/1950). Gestalt theory. In W. D. Ellis (Eds.), *A sourcebook of Gestalt psychology* (pp. 1–10). New York: Humanities Press.

Wolfe, J. M. (1999). Inattentional amnesia. In V. Coltheart (Ed.), *Fleeting memories.* Cambridge, MA: MIT Press.

Womelsdorf, T., & Fries, P. (2007). The role of neuronal synchronization in selective attention. *Current Opinion in Neurobiology, 17*(2), 154–160.

Zipser, K., Lamme, V. A. F., & Schiller, P. H. (1996). Contextual modulation in primary visual cortex. *Journal of Neuroscience, 16,* 7376–7389.

Printed and bound by CPI Group (UK) Ltd, Croydon, CR0 4YY

23/10/2024

01777678-0020